Causation and Delay
in Construction Disputes

Causation and Delay in Construction Disputes

Nicholas J. Carnell LLB (Hons), FCIArb, Solicitor

Blackwell Science

The right of the Author to be identified as the
Author of this Work has been asserted in
accordance with the Copyright, Designs and
Patents Act 1988.

First published 2000

Set in 10.5/13pt Palatino
by DP Photosetting, Aylesbury, Bucks
Printed and bound in Great Britain by
the University Press, Cambridge

DISTRIBUTORS

  Marston Book Services Ltd
  PO Box 269
  Abingdon
  Oxon OX14 4YN
  (*Orders:* Tel: 01235 465500
         Fax: 01235 465555)

USA
  Blackwell Science, Inc.
  Commerce Place
  350 Main Street
  Malden, MA 02148 5018
  (*Orders:* Tel: 800 759 6102
            781 388 8250
         Fax: 781 388 8255)

Canada
  Login Brothers Book Company
  324 Saulteaux Crescent
  Winnipeg, Manitoba R3J 3T2
  (*Orders:* Tel: 204 837-2987
         Fax: 204 837-3116)

Australia
  Blackwell Science Pty Ltd
  54 University Street
  Carlton, Victoria 3053
  (*Orders:* Tel: 03 9347 0300
         Fax: 03 9347 5001)

A catalogue record for this title is available from
the British Library

ISBN 0-632-03971-X

Library of Congress
Cataloging-in-Publication Data
is available

For further information on
Blackwell Science, visit our website:
www.blackwell-science.com

Contents

Contents

Contents

Preface

Delay claims are perhaps the most common form of construction dispute. Few people in the construction industry can claim never to have known a project which ran late. However, it is also a subject which is still widely misunderstood. Although it will soon be ten years since the Privy Council's restatement of the law in *Wharf Properties* v. *Eric Cumine Associates*, the terms 'cause and effect' and 'rolled up claim' are often treated as if some magical significance attaches to them. Too often the result is that, in addition to being faced with a project which has gone badly, the contractor or developer is confronted by what appears to be a bewildering set of obstacles placed between him and establishing his entitlement.

This book is not aimed at lawyers; the intention is to provide a reference guide for construction professionals. The book has attempted to look at some of the practical considerations which can lead to problems on any project, particularly record keeping and notices, followed by consideration of some of the more frequently encountered contractual issues, such as the entitlement to rely on a programme and the circumstances in which time might be said to become at large. This precedes an analysis of the principal cases concerned with proving delay claims, starting, of course, with *Wharf Properties*. There follows a brief consideration of the techniques involved in network planning and the use of the critical path analysis to prove delays. The final two chapters are concerned with claim presentation and the various forms of dispute resolution which are available.

Since this is intended to be a practical guide, and the scope of the subject is potentially enormous, and since a number of the subjects addressed would merit a work in their own right, in many instances topics are given a fairly brief introduction which it is hoped will lead the interested reader on to more detailed reading.

Perhaps the single most important development since this book was conceived is the introduction and development of adjudication as a quick means of resolving disputes. The impact of adjudication on the various subjects covered is addressed both by reference to the individual matters covered and in Chapter 10. The view expressed is generally that adjudication will certainly serve to compress the timescale within which disputes are resolved and lessen some of the procedural requirements often regarded as making litigation and arbitration unnecessarily complex, but it is unlikely, at least in the foreseeable future, to bring about a significant relaxation in the standards of proof that are likely to be required.

The growing use of adjudication, however, is perhaps only the most obvious sign of the changed climate in construction over recent years. Chapters 1 and 10 both comment on the call by both Latham and Egan for construction to be carried out in a climate in which disputes are the exception rather than the norm. Wisely, however, Sir Michael Latham concedes that disputes are part of commercial life and so calls for dispute resolution procedures which allow the prompt disposal of disputes. It is implicit that such disputes will be capable of resolution more swiftly and painlessly where both parties understand and acknowledge the same set of rules, even if they do not agree with one another on the facts. Far too many delay claims come to resemble trench warfare because, in addition to disagreeing with one another over the causes of delay, the parties also disagree about what needs to be proved and how. It is hoped that this book will in a modest way assist in the growth of such understanding.

The other major reform of recent years has been the new Civil Procedure Rules, the wholesale rewriting of the rules of civil litigation in the wake of Lord Woolf's 1996 Report *Access to Justice*. Allied to the 1996 Arbitration Act and the growth of adjudication, the whole basis on which disputes are resolved has undergone its most radical overhaul since the Judicature Acts at the end of the 19th century. At the time when the manuscript for this book was completed, the Protocol for the adaptation of the Civil Procedure Rules for use specifically in the construction disputes dealt with by the Technology & Construction Court was still in draft form. The discussions as to the final form which this Protocol will take have, it is understood, been both lively and far reaching. The decision was therefore taken, after a good deal of thought, to omit specific discussion of that document. However, it appears clear that, whatever form the Protocol eventually takes, it will require the parties to a dispute to meet prior to the formal proceedings with a view to identifying not only the areas of dispute, but also the methods by which they are to be addressed.

As far as possible, references to standard form building contracts have been kept to a minimum. The great majority of references are to JCT 98, the Second Edition of the Engineering and Construction Contract (while maintaining the NEC abbreviation) and the ICE 7th Edition. It is intended that the law is as stated at 30 September 1999.

Author's note

One of the very best legal text books, *Dicey & Morris on the Conflict of Laws*, also boasts what is certainly the best Author's Note. Professor Morris says, in effect, that law books are much like babies – great fun to conceive but thereafter colossally hard work. I can only add that, in the course of writing, my admiration for those who have undertaken more substantial and learned works than this has increased many times over. It is also fitting in any work concerned with construction disputes to acknowledge my debt of gratitude to that large body of people, largely unsung and frequently maligned, who make up the construction industry past and present. It has been my great good fortune over the past 15 years to work with some of these people. More than any reported case or law book, their legacies are the buildings and works of civil engineering which they have constructed.

Nicholas J. Carnell
May 2000

Acknowledgements

Firstly and of course, I am indebted to my wife, Debbie, for her encouragement, patience and support during the writing of this book, and for the seemingly endless sacrifice of our dining room table to this task. Secondly, to my secretary, Pamela Sylvane-Addo, for converting my two fingered typing into something presentable. Thirdly, to Julia Burden at Blackwell Science for her patience, firmness, charm and occasional credulity in dealing with an author whose approach to time limits would have driven lesser mortals to despair.

Thanks are also due to my colleagues in the S J Berwin & Co Construction Group, particularly Ian Insley, Jonathan Dockney, Anna Sharvatt, Louis Flannery and Martin Scott, who have offered numerous helpful suggestions, which I duly acknowledge whether I have incorporated them or not. Needless to say, the views expressed are my own and I take full responsibility for them.

List of Abbreviations

ADR   alternative dispute resolution
CPA   critical path analysis
EET   earliest event time
EFT   earliest finishing time
EST   earliest starting time
ICE   Institution of Civil Engineers
IFC 98   JCT Intermediate Form of Building Contract 1998
IMechE   Institution of Mechanical Engineers
JCT   Joint Contracts Tribunal
JCT 98   JCT Standard Form of Building Contract 1998
LET   latest event time
LFT   latest finishing time
LST   latest starting time
MW 98   JCT Agreement for Minor Building Works 1998
NEC   New Engineering Contract
PC   prime cost
PERT   programme evaluation and review technique
PNT   project network techniques
PQS   professional quantity surveyor
SBIM   School of Business and Industrial Management
TPT   total project time
WCD   With Contractor's Design
WCD 98   JCT Standard Form of Contract With Contractor's
          Design 1998

CHAPTER ONE
TIME IS MONEY

1.1 Introduction

This book is intended to act as a guide through the construction process for those engaged on behalf of both employers and contractors, and to provide an aid in avoiding delays and also in coping with them when they do arise. Quite deliberately the subject of quantification of claims has not been attempted. This is because it has already been dealt with by others in a manner which the author could not possibly hope to emulate, to say nothing of the fact that the present task is already a substantial one.

No two construction projects are alike; accordingly, no two delay claims will ever have identical ingredients. Even the simplest series of modular buildings will be erected on different pieces of land or at different times or by different people. Unlike manufacturing industry, construction is not primarily concerned with the repetition of a series of processes but with a succession of one-off projects. Hence, as the complexity of the works increases, so does the number of variables, and, of course, the range of things which can go wrong.

The cost of a project will be determined by an equation which balances time, materials and labour against the conditions under which the works are to be executed and the requirements of the person for whom the works are being carried out. Planning a project is concerned with determining how many men with what equipment will take how long to carry out what work on a particular site. The project will be costed by determining the quantities of these components which will be required to complete the required work, and, where one side of the equation undergoes a significant alteration, claims will frequently follow.

The origins of modern construction and civil engineering planning lie in the canal boom of the late 18th century. Prior to that time most employers had hired individual craftsmen and labourers. The canal age saw the workforce grouped into gangs under the ultimate control of the engineer. The requirement that each new 'navigation' should have an enabling Act of Parliament to allow the formation of a new joint stock company meant that budgets in the form of the company's share capital were fixed in advance by reference to the anticipated cost of the works. The results were projects which would be instantly recognisable, particularly to those involved in construction management. Equally familiar were the delays and increases in costs which bedevilled many of these projects.[1.1]

Then, as now, a great deal of energy was expended attempting to plan projects in such a way that completion took place on time and within budget. When it did not, claims resulted, and again these would be familiar to today's contractors.

1.2   *An outline of the battlefield – looking forward*

Where one of the key resources is significantly altered the result will generally be either delay to the works or the need for acceleration. To understand either we will have to give brief consideration to the planning of the project. The first question facing the planners of every job is 'How do we propose getting this project from inception to completion in accordance with the programme and budget?' The important point to realise is that although no two jobs are ever exactly the same, and thus the number of potential things which can go wrong is infinite, these problems fall into a series of broad categories. These can be anticipated, and steps taken to guard against them.

Indeed, these broad categories can really be grouped into two headings – those which have their origins in the planning of the project and those which are caused by problems during the construction of the works.

Planning the project

Accordingly, the starting point in understanding delay claims is the period before work has started on site and appreciation of the following matters.

- The importance of planning the works properly, which above all means within time and budget constraints that are actually capable of being fulfilled. This obviously starts with the employer producing a scheme for the procurement of the works which is possible within these parameters.
- The role and preparation of programmes, histograms and resource schedules by the contractor to enable the requirements of the employer to be fulfilled.
- The part to be played by critical path analysis in planning the works.
- The parties' contractual obligations and entitlements, particularly in the principal standard form building and civil engineering contracts and sub-contracts, and especially those provisions regulating time.

During the construction

Only then is it appropriate to look at the matters which actually cause delay during the works themselves. These will typically be one of two types.

Contractor's responsibility

Those which arise due to a failure on the part of the contractor. Some of these will have their origins in a failure properly to carry out the planning stages of the works, others will be due to an inability to perform in the manner agreed in the contract.

Employer's responsibility or neutral events

Those caused through an act or omission of the employer or his team or by a matter which does not arise through the fault of the contractor. These will also be governed by the contract conditions. A useful list of these is provided in Clause 25.4 of JCT 98 and includes:

- *force majeure*
- exceptionally adverse weather conditions
- clause 22 perils (flood and the like)

- civil commotion, strike or lock out
- compliance with architect's instructions
- non-receipt of essential information
- delays by nominated suppliers or sub-contractors, artisans and tradesmen
- Government action restricting labour or materials
- delays by statutory undertakers
- delays in giving access to the works.

Dealing with the claim

That done, attention turns to the task facing surveyors, lawyers or claims consultants, typically, and perhaps unfortunately, coming to the project after the delaying events have occurred and the project is significantly late (matters considered in detail in Chapters 7 and 8). Invariably their task is to produce or rebut a claim, seeking to assert that delays are the fault of someone other than their client. This will involve analysis of the methods frequently employed in this exercise and the problems with each such approach. This cannot be done without then considering the guidance and sometimes hindrance provided by courts. It is also apt to bear in mind, even at an early stage, the levels of proof required by the courts should the parties fail to reconcile their differences.

It is then appropriate to look in Chapter 9 at some of the ways in which claim preparation can be improved. The objective is to identify the steps to be taken in producing claims which will achieve their forensic objective – proving why delays occurred. Necessarily this will involve a brief guide to 'the Black Museum' – those claims which have gone badly wrong and where shortcomings have been exposed and highlighted by the court – since the invariable truth is that it is easier to determine what can and should be done with the benefit of hindsight and from others' experience.

It is only fair at this point to declare an interest. My own experience of delay claims and their causes comes from advising those in the construction industry, most frequently in circumstances where the delay is already a fact and the issue is how it occurred and whose fault this is. The greater part of this book is written from the perspective that the parties to a construction project have their rights and obligations mapped out by the contract by which they have agreed to be bound. Delays and claims result from matters which mean that the works are not carried out precisely as envisaged in that contract. Accordingly, avoiding delays is crucially concerned not only with good practice during the planning and execution of the works but also with proper operation of the contract machinery. Similarly, successfully mounting or defending a claim is largely an exercise in understanding and enforcing rights and duties contained within that contractual framework.

Old and new approaches

In many respects this is a view which is as old as contracting itself. However, and almost by definition, it is a confrontational approach in which the contract serves, as the heading of this section suggests, to outline the battlefield. Nevertheless, while co-operation is not always seen as the way to achieve the best results, in reality it will usually pay dividends. A very large part of this book is therefore concerned with discussing techniques which can be used to obtain the best results available under the contract.

This view has been questioned by three important developments. These were:

- the publication of the New Engineering Contract (NEC) – now the Engineering and Construction Contract,
- the release of *Constructing the Team* by Sir Michael Latham, the final report of the government/industry review of procurement and contractual arrangements in the UK construction industry, and
- the Housing Grants, Construction and Regeneration Act 1996.

All three were keenly anticipated and, since publication, have provoked lively debate. The impact of each will be considered in subsequent chapters. For present purposes, it is important to note that while making the approach from slightly different angles, each addresses the industry from a novel perspective. This, essentially, is that the problems of the construction industry emanate from contractual relationships which provoke conflict rather than consensus.

The NEC approach is to impose a duty of good faith, while the Latham Report calls for a change in attitude – to promote co-operation rather than conflict between the parties. The Housing Grants, Construction and Regeneration Act 1996 attempts to provide for fairer dealings between parties to construction contracts by requiring interim payments to be made, limiting the right of set-off, outlawing 'pay when paid' clauses and, most importantly for present purposes, providing a statutory right to have disputes dealt with quickly and efficiently by way of adjudication. (Although the right to adjudication does not apply to contracts entered into before 1 May 1998). While adjudication will significantly reduce the time period required to deal with disputes, experience suggests that this will not be at the expense of a lowering of the required standard of proof.

While the overall objectives of each development deserve the very highest praise, the commentators have given a reception which has been decidedly mixed.[1,2] It is nevertheless interesting that, when first published, the drafting of the NEC was criticised as being change for its own sake, and certainly, some of the drafting of the first edition was clumsy and imprecise; but many of these difficulties have been addressed in the second edition. More importantly, as the NEC becomes more commonly used, it seems likely that many of those criticisms will be 'worked out'.

The final paragraph of this opening chapter offers an apology in respect of two comments which will certainly be levelled at this book. The first is that it comprises a counsel of perfection; that this is all very well but it will involve so much care on the part of all those involved in the building process as to prevent any project from getting past the planning stage. The second is that it is obviously easy for a lawyer to offer suggestions on how best to deal with delay claims from a perspective which is necessarily concerned with generalising problems. Both are valid criticisms. However, the response to each is basically the same. This is not a complicated book. The view of the majority of specialist lawyers engaged in this field is that most problems arise from simple and, in the main, avoidable events. The steps required to avoid problems are frequently no more arduous than appreciating what precisely the contract or even just good sense dictate. In the main, this involves developing good habits. The irony is that the provisions of the relevant British Standard BS 5750 require precisely this. While it is obviously appreciated that the majority of those employed in the industry are exceptionally busy, the issue is one of suggesting how the limited time available can best be spent. These are universally applicable considerations.

CHAPTER TWO
PLANNING THE PROJECT

2.1   Allocating risk

Choice of contract

Every project starts with a decision by the employer to carry out certain works. After determining what he wants to build the next decision, and possibly one of the most crucial in the whole project, is to decide upon the contractual regime according to which the works are to be executed. Virtually everything which follows will depend upon this decision. This is an obvious point but one which should not be ignored. To give an obvious example; the choice between design and build and traditional contracts involves deciding between two wholly different ways of allocating risk. Whether the employer retains control over, and therefore responsibility for, the design of the works or whether he delegates this task to the contractor, protecting his own position by a design warranty of the sort found in clause 2.5.1. of the JCT Standard Form of Building Contract With Contractor's Design.[2.1] The choice of permutations and the range of available forms of contract is enormous. Some of the questions which face the potential employer in picking an appropriate regime for the works are dealt with below.

The range of choices has expanded considerably over the past decade, not only in terms of the different types of standard forms now published but also in relation to the number of ways in which parties attempt to modify those standard forms. Depending upon precisely which version is selected the unamended JCT 98 standard form is just over 60 pages long. It is not unusual to see this augmented by amendments proposed by one or other of the parties which can add the same length again. The authors of these amendments are generally either lawyers or project managers. Except where the works have some truly extraordinary feature, this is a practice which has been condemned in all quarters as a fairly naked attempt to secure an advantage and shift the balance of risk in the standard form.

Figure 2.1 shows the sort of network of relationships which may be involved and the contractual links between the parties. Each of these relationships involves the allocation of risk between the parties. The importance to the employer and his professional advisors of choosing the appropriate form of contract, whether it is a standard form or a tailor-made document, cannot be overstressed. The wrong choice can have serious consequences later in the con-

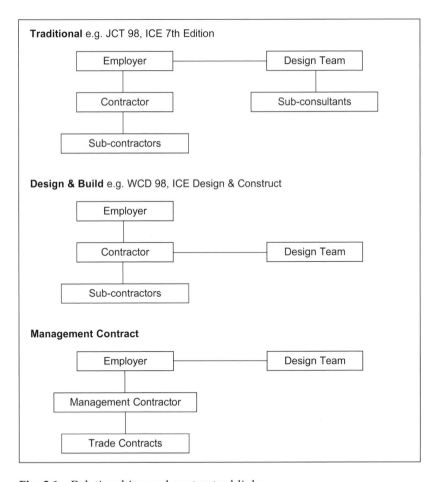

Fig. 2.1 Relationships and contractual links.

tract. An illustration of this is provided by comparison between two forms which are often used for similar works, namely the ICE 7th Edition and the IMechE Model Form. Both have been widely used by local authorities, particularly in connection with street lighting contracts. The former provides by clause 12 that where unforeseen ground conditions are encountered this may entitle the contractor to relief. By contrast, clauses 2 and 11 of the IMechE Form make the contractor responsible for determining ground conditions, and the right to claim that ground conditions are unforeseeable is thus limited. Hence, the employer faces a clear choice between the two forms and the decision will depend in large part upon the employer's perception of which regime is most likely to suit the conditions on a particular job.

Initial questions

As Fig. 2.1 shows, the different standard forms and different contractual schemes impose completely different sets of relationships. In a recent paper presented to the King's College Centre for Construction Law and Management,[2.2] Richard Winward asked the question whether construction contracts actually benefit the parties. His starting point was to refer back to the report by Sir Harold Banwell in 1964[2.3] where the problems in the construction industry at that time were attributed to the failure to use and comply with standard form contracts. The industry has moved on but the possibility still exists for the parties to select a particular standard form for a project which does not suit the objectives they are attempting to achieve. The choice of the correct form is obviously dependent upon the precise circumstances pertaining to the project, but, without producing a comprehensive list, the following questions may be worth repeating.

(1) Does this project actually merit the use of the Minor Works Form or Intermediate Form instead of their 'grown up' relatives?
(2) Do the number of PC or prime cost items mean that it would be more desirable to use a re-measurable form?
(3) Is this a contract which merits the production of full scale bills of quantities or is something more standardised going to suffice?

(4) Do we really need a design team to act on behalf of the employer or would it be more sensible to produce a conceptual design and then delegate the detailed design work to the contractor? Do we therefore propose to novate or assign the authors of the original design to the contractor?

(5) Is the complexity of the project such that we can entrust co-ordination of the works to the contractor or the architect, or would it be better to have this task performed by a specialist project manager?

(6) Would it be sensible to do away with the idea of the main contractor altogether and acknowledge that the works are so complex that what we actually need is a management con-tractor whose job is to manage the works of the various package contractors who will build the project?

(7) Alternatively, is this the sort of job where we are building something which in itself is fairly straightforward but where the circumstances or conditions in which it is being built are such that the price of the works may vary considerably depending upon how we progress?

(8) At the opposite end of the scale, is this a case where we simply pay the contractor a sum of money and at the end of the contract period take delivery of a completed project?

Identifying responsibilities

It will be apparent from Fig. 2.1 that the choice of contract will be instrumental in determining where risks lie. How this affects the timing of the works will be considered in the next section. The ascertainment of delay, its calculation, causes, and most impor-tantly responsibility are all related back to what the contract actu-ally says. Many claims are badly produced because the claimant fails properly to consider the precise nature of the obligations actually imposed.[2.4]

It almost speaks for itself that the wrong choice of contract has potentially serious consequences. This is illustrated by taking a simple situation and comparing the different risks. Take as an example the construction of a warehouse where problems occur with the installation of the exterior cladding, caused by either faulty design of the fixing system or poor workmanship, resulting in

11

instructions changing the fixing to a different method. The following permutations are possible, and for these purposes, it is probably sufficient to compare the traditional regime contained in JCT 98 with the design and build method in WCD 98.

Design

The scheme of WCD 98 is that the employer by his Requirements says what he wants to have built and the contractor by his Proposals tells him how he proposes to do this. Clause 2.1 obliges the contractor to carry out the works in accordance with among other things the Employer's Requirements and the Contractor's Proposals. If the problem is one of defective design, under clause 2.5.1, to the extent that the works are designed by the contractor and form part of the Contractor's Proposals, the contractor owes the same duty as if he were an architect employed under a separate contract. This is a cumbersome formula but its meaning is clear enough – the contractor must exercise reasonable skill and care in the design work.

By contrast, under the terms of clause 2.1 of JCT 98, the contractor's obligations are limited to the execution of the works. He has not prepared Contractor's Proposals and design is primarily the responsibility of the architect appointed by the employer (see Fig. 2.1).

Workmanship

Under both régimes, this is the contractor's risk.

Instructions

Under JCT 98, an instruction altering the design of the works can be made under the terms of clause 14.1.1, and the contractor is entitled to have the varied work needed to implement that changed design valued under the rules prescribed in clause 14.5. By contrast, WCD 98 makes it clear that a Change in the Employer's Requirements can be made under clause 12.1 and will be valued in accordance with clause 12.5, but by necessary implication, and by operation of clause 8.1.2, if the works do not accord with the Employer's Requirements or are not of a standard appropriate to the works necessitating design modifications, the risk rests with the contractor.

Extensions of time and loss and expense

Under JCT 98, instructions requiring a variation of the type described in the previous paragraph will be a relevant event for the purposes of both clauses 25.3 and 26.5 and will entitle the contractor to both extensions of time and appropriate loss and expense. In WCD 98, the fact that these matters are at the risk of the contractor will obviously disqualify him from recovering extensions of time or loss and expense, except to the extent that the instructions or variations come about as a result of a Change in the Employer's Requirements or the eradication of an inconsistency in the Employer's Requirements.

The point is one which cannot be overstressed – the starting point for any analysis of delay or entitlement to extensions is the contract. If the contractor has undertaken a particular risk, he cannot complain if it causes him delay.

2.2   *Timing obligations*

The choice of contract will also govern the parties' timing obligations. The question is not only 'How long do we have to complete the various activities comprising the works, and the works themselves?' but also 'If we do not complete by these dates, which events will influence whether we can claim more time and obtain some recompense for this?' Taking the example of risk allocation above it will be apparent that the answer will lie in whether certain functions lie within the range of risks retained by the employer or whether they fall under the control of the contractor and are items which will have been taken into account in calculating the price.

Timing is crucial in building and civil engineering contracts because of the extent to which the price of the works is dictated by the time related costs of plant, labour and overheads. This is obvious if one considers and compares the lesser degree to which the price of a car will relate to the time based components in its price. Of equal importance is the fact that the construction industry, as we all know, is concerned with producing a never ending series of prototypes. To that extent, the amount of time required to complete a project is always likely to involve a degree of guesswork. A feature of the tendering process is the calculated gamble by the contractor to determine how much of any particular variable resource is necessary to deliver the project within the time parameters while producing a competitive price. This is frequently identified as the source of disputes. In a depressed market such as that which characterised the early 1990s, the result was that market forces led to many jobs being priced at levels which would not ordinarily permit the contractor to cover his costs, let alone make a profit if he devoted sufficient resources to execute the works in accordance with the contract.

For the present it is sufficient to make two comments about this policy of buying work. The first is that, if the price for the works is only achievable by paring resources to a bare minimum, events which the contractor might otherwise have taken in his stride will impact more seriously upon his ability to proceed in accordance with the contract. Hence delays are more likely and will be greater when they do occur. The second is that in these circumstances the only way in which the contractor is likely to cover costs is through variations and claims. For an employer, therefore, obtaining a rock bottom price is potentially a dangerous business. These are precisely the sort of contracts with the in-built capacity to go wrong.[2.5] Both of these points will be considered further in the next section in relation to the planning of the works.

The basic position – no time obligations

It is necessary to approach timing obligations from first principles. In the very simplest arrangements a client will simply ask the contractor to carry out work, which may or may not be defined. In these circumstances, there is no contract in the legal sense. The work is carried out on a day to day basis. The client has no entitlement to require the contractor to carry out any specific quantity of work. The contractor can down tools whenever he chooses. In return for the work which is done the contractor is entitled to payment. The method to be adopted in determining the payment to which he is entitled is the subject matter in its own right of several textbooks. At its very simplest, payment is probably to be calculated by reference to costs and a reasonable addition in respect of profit and overheads.

What is clear is that, in this situation, the client cannot require that the works be carried out in any order or that they are finished by a particular date. The client has no recompense if the contractor takes longer over the works than the client thinks he should. The principle is clear: obligations in respect of time come from the contract, from what the parties agree, or are taken to have agreed.[2.6]

The next stage – reasonable time

Moving up the evolutionary chain, in simple contractual arrangements the contractor agrees to carry out certain works in return for payment of a particular sum. If the parties leave open the timing of the works, the contractor will have a reasonable time in which to execute them. This expression has proved difficult to define for the simple reason that 'reasonableness' will obviously be dictated by the circumstances which prevail on a particular project. A helpful example is provided by the judgment of Mr Justice Goff in *British Steel Corporation* v. *Cleveland Bridge & Engineering Co Ltd*, (1981) 24 BLR 94. The parties accepted that the proper test was to consider the circumstances which existed when the works were carried out, excluding those matters which were under the exclusive control of the party carrying out the works. The judge proceeded to look at the various constituent causes of delay and assessed how long should reasonably have been taken to carry out the various activities, compared against the actual time taken. He based his view upon the evidence of the parties. The judge conceded that this was necessarily a rough and ready approach. It is also important to remember that the judge had previously held that in fact the plaintiffs were under no obligation to deliver the subject matter of the claim within a reasonable time, and hence his comments, while useful, are not to be regarded as comprising binding authority for any proposition of law.

Prior to this case, in an unreported Court of Appeal case, *Sanders and Forster* v. *A Monk & Co. Ltd*, it had been held that,

'What is reasonable depends upon all the circumstances of the case, including the conditions operating during the period when the work is being done. It would follow that as to whether a reasonable time has been taken up in the doing of the work

15

cannot be decided in advance; it can only be decided after the work has been done.'

Two points follow from this:

- The impossibility of defining the term 'reasonable time' except by reference to the prevailing facts means that if disputes arise it is difficult to see an easy way to resolve them other than by proceedings.
- It is far easier to allege that the works were completed within a reasonable time than it is to disprove it. The various cases on the question show this.

In *Pantland Hick* v. *Raymond & Reid* [1893] AC 22, Lord Watson in the House of Lords held that a person obliged to complete within a reasonable time does so, 'notwithstanding protracted delay, so long as the delay is attributable to causes beyond his control, and he has acted neither negligently nor unreasonably'. The clear inference to be drawn from this is that a fair degree of latitude will be given to the person alleging an entitlement to a reasonable time.

In the absence of a clear indication to the contrary, without some plain evidence on the subject, courts and arbitrators will be reluctant to write a timing obligation into the contract where none otherwise exists.

Development of timing obligations – duties

The situation described above is unsatisfactory. In all but the simplest contracts, timing is critical to both parties. Historically, as building contracts evolved, the obligation to carry out work was coupled to a fixed period within which the works were to be executed. As a matter of general law, if the contractor defaults the employer can claim general damages in accordance with the ordinary principles of damages for breach of contract. However this left uncertain the extent of the consequences which would face the contractor if he delayed.

Predictably, the parties to the contract attempted from the Middle Ages onwards to produce a mechanism for fixing this in advance of the works. Sir Michael Latham refers to James Nisbet's fascinatingly gruesome *Fair and Reasonable – Building Contracts since 1550* where delay by contractors is said to have led to punishments ranging from excommunication to imprisonment and confiscation of goods. Employers rapidly appreciated that better results were achieved by threatening the contractor's wallet as opposed to his immortal soul. Hence the development of penalties – or as they subsequently became, liquidated and ascertained damages.

Development of timing obligations – rights

The above, of course, fails to address the question of extensions of time. The point was considered in *Holme* v. *Guppy* (1838) 3 M & W 387. In that case, the contractor had a period of four and a half months in which to complete the works, failing which he became liable to pay liquidated damages. In the event, the contractor was unable to complete as a result of the acts of prevention of the employer. The employer sued for liquidated damages as a result of the contractor's delay and, dismissing the action, Baron Park said that 'if a party be prevented by the refusal of the other contracting party from completing the work until the time limited, he is not liable in law for that default'.

This is the principle that a party cannot rely upon his own default. That, however, left open the question of the contractor's obligations in these circumstances. The judge approached the matter from a common sense perspective holding that:

- the contractors were excused from performing the works within the original four and a half months, and
- there was nothing to indicate that they had agreed to any other timing obligation, and therefore,
- any timing obligation in the original contract must be taken to have broken down.

Thus, time was at large i.e. the only obligation was to complete in a time reasonable in all the circumstances, and the employer was not entitled to recover anything in respect of liquidated and ascertained damages.

Matters were developed somewhat in *Roberts* v. *Bury Commissioners* (1870) LR 5 CP 310. The dispute has an almost contemporary

feel to it – in essence it concerned a claim by employers for liquidated damages, met by an allegation by the contractor that he had
not been supplied with essential information within sufficient time
to enable him to complete within the contract period. The court's
finding is put slightly differently from the way in which Baron Park
had put matters in *Holme* v. *Guppy* namely that

> 'The contractor also from the nature of the works could not begin
> the work until the commissioners and their architect had sup-
> plied plans and set out the land and given the necessary parti-
> culars; and therefore in the absence of any express stipulation on
> the subject, there would be an implied contract on the part of the
> commissioners to do their part within a reasonable time.'

The important point is not the reference to implied contracts, but the
notion that timing obligations were to be seen as a two-way street,
conferring rights as well as duties on the contractor.

This point was refined in *Wells* v. *Army and Navy Co-Operative
Stores* (1902) 2 HBC 346. This case is still generally regarded as
setting out the law as it presently stands. Once again it involved the
assertion that the contractor had been prevented from carrying out
the works within the agreed contract period as a result of the
employer's failure to supply details and drawings in proper time.
The classic statement of Lord Justice Vaughan Williams in the Court
of Appeal is worth setting out in full

> '[I]n the contract one finds time limited within which the builder
> is to do the work. That means not only that he is to do it within
> that time but it means also that he is to have that time within
> which to do it ... in my mind that limitation of time is intended
> not only as an obligation, but as a benefit to the builder.... In my
> judgment where you have a time clause and a penalty clause (as I
> see it) it is always implied in such clauses that the penalties are
> only to apply if the builder has, as far as the building owner is
> concerned and his conduct is concerned, that time accorded to
> him for the execution of the works which the contract con-
> templates he should have.'

In both *Holme* v. *Guppy* and *Roberts* v. *Bury Commissioners* the
contracts lacked a mechanism for extending time. In *Wells*, although

the contract did contain such a provision, the court held that the machinery was insufficiently wide to permit the granting of an extension of time. Hence, time was once again set at large. Modern contracts contain significantly more sophisticated mechanisms for determining the extensions of time which should be granted. However, the importance of these ancient cases is to set in stone the principle that timing requirements confer both rights and duties.

It would be unwise to assume that these decisions give support to a general line of argument allowing timing obligations to be set aside in the sort of circumstances which occurred in those cases. Instead it is worth summarising the principles for which they stand:

(1) Where a contract contains a provision which obliges a party to do specific things within a particular period of time, that party is also entitled to have that period in which to do the work.
(2) Hence, if the party for whom he is carrying out that work conducts himself in such a way as to deny that period to the contractor, he cannot then rely upon an entitlement to liquidated and ascertained damages.
(3) This means that an obligation to carry out work within a specific period is not to be seen as an unqualified or absolute obligation. Instead it is a duty backed by a right.

A further and more general point occurs. That is the idea that delay claims all progress:

- from consideration of what obligations are imposed by the contract,
- to analysis of the events which gave rise to the problems encountered by the parties,
- to determination of what resulted from those events.

2.3 Planning tools

At its simplest a building contract generally comprises an obligation to carry out specific work over a particular period in return for a fixed reward. It will be apparent from the first part of this chapter that the provisions of the various standard form contracts are designed to set out a variety of regimes within which this simple objective can be achieved. The history of the construction industry over the past decade shows that there has been a widespread movement towards increasing the sophistication with which projects are planned. The thinking is very simple; if everything which possibly can be planned is addressed before the works are started, there will be proportionately less to go wrong. Figure. 2.2 provides a quick checklist of the sort of planning tools which can be utilised and the purpose which they attempt to serve.

Labour Histogram	To determine level of labour resource required at particular points during execution of works, and factors to be costed in tender, and to ensure availability of sufficient labour to meet progress requirements.
Material Histogram	As above.
Access dates required Schedule	Dates access required to particular areas of work to maintain progress/meet programme.
Information Release Schedule	Dates when particular drawings or information required to progress works in accordance with the programme.

Programme
  Master Programme
 	
  Short Term Programmes
  Network Analysis
	To plan order for execution of works, including:
	• The sequence for construction of the whole of the works;
	• For construction of particular activities or during particular periods;
	• To demonstrate dependencies between particular activities

Design Freeze Dates	Date(s) by which all design information shall have been provided in final form.

Fig. 2.2 Planning tools.

Amending standard form contracts

Of course it will never be possible to guard against everything which can possibly happen. As we have seen a contract is no more than a way of allocating risk and how the parties go about this is ultimately up to their relative negotiating skills and bargaining power. What is significant is the premise that the way in which they distribute risk is much less important than the fact that there is certainty from the outset.[2.7]

This is important. Experience tends to suggest that one of the most common causes of disputes is a failure on the part of the contracting parties to appreciate what they have actually let themselves in for. A dispute develops from a mistaken belief that the contract can be operated in a manner other than that which the agreement actually permits. Obviously any businessman will attempt to secure an advantage from a transaction but if the parties do not understand their respective rights and obligations it will be less likely that they will be able to avoid dissension arising. It is also frankly unhelpful to concentrate too much upon 'fairness' which after all is an abstract concept, and what is fair for one party may be unfair for another. Common sense suggests that fairness is much less important than certainty – if both sides understand exactly how the relationship is to be regulated, it is more likely that they will attempt to make the best of the contract, whether or not it could objectively be regarded as fair.

However, properly approached, there is scope for the standard forms to be improved. It follows that this is best achieved where any amendments are designed to increase the level of certainty within which the parties are operating rather than attempting to alter the way in which the contract approaches the allocation of risk.

Programmes

The most obvious example of this comes in relation to defining the parties' timing obligations with greater precision. Of course, with the exception of the ICE and NEC forms, none of the most commonly used standard forms impose an obligation on the parties to produce, still less to agree a programme for the execution of the works. Hence, if we take the unamended JCT 98 form as providing an example:

- The First Recital provides that the employer has produced drawings and, as appropriate, bills of quantities or approximate quantities showing and describing the works or a reasonably accurate forecast of the works.
- The Second Recital provides that the contractor has priced the works, where applicable by reference to the bills or approximate bills.

21

- Articles 1 and 2 provide that that the contractor will carry out the works in accordance with the contract documents and in exchange for payment of the sum or sums which become payable in accordance with the provisions of the contract.
- Clause 2.1 of the contract conditions provides that the works will be carried out in accordance with the contract documents which include the drawings, bills (depending upon the version of the contract being used), articles of agreement and Appendix.
- Clause 5.3.1.2. provides that among the documents to be provided by the contractor shall be two copies of his master programme for the execution of the works. Clause 5.3.2 provides that nothing in the documents provided under clause 5.3.1, nor in the master programme, shall impose a greater obligation than that imposed by the contract documents. This is the only reference in the contract to a programme; the expression is not defined in clause 1, nor is there any requirement that the programme should actually be produced.
- The Appendix provides dates for possession and completion of the works.

By contrast, clause 14 of ICE 7th Edition requires the production of a programme, while clause 31.1 of the NEC 2nd Edition requires both the production of a programme and the subsequent measurement of progress by reference to that programme, with the programme being up-dated to take account of progress.

The potential problem with JCT 98 is obvious. Provided that the contractor starts and finishes on the prescribed dates he can do the works in whatever manner or sequence he pleases. This was in fact identified in *Wells* v. *Army and Navy Co-Operative Stores* by Mr Justice Wright the judge at first instance, who said

> 'The plaintiffs must within reasonable limits be allowed to decide for themselves at what time they are to be supplied with detail.'

The concern which faces employers is that in complex projects there must be some way of plotting, and indeed regulating, the progress of the works. While of course the overwhelming majority of contractors will produce a programme, whether or not there is any obligation to do so, that programme is no more than a convenient aid for the contractor – failure to observe its provisions cannot justify a complaint on the part of the employer. The other side of the coin is that if delay results it is much harder to get to the root causes. Delay is generally calculated by reference to the completion date, but it will be obvious from the cases referred to in the previous section (and this is common to all of the later cases) that to succeed the parties must actually come up with some credible account as to why the completion date was missed. Without a pre-set programme this is very difficult and the sad fact is that disputes become almost inevitable. The production of a programme does at least serve to provide the parties with a yardstick by which they can assess when certain events should have happened and when they actually took place.

Hence good sense has come to suggest that the practice of producing a programme as a construction aid should be developed to provide something rather closer to a formalised route map for the works. It is now commonplace to see clauses in the preliminaries to the bills, or the equivalent section in the Employer's Requirements (WCD 98), imposing an obligation upon the contractor to produce a programme in a particular form by a given date, and providing specific dates for the achievement of certain events. Thus, the apparent gap in clause 5.3 of JCT 98 is filled. It is becoming almost as common to see this requirement followed by stipulations as to the production of a critical path analysis marked on the programme to demonstrate the contractor's proposed critical path for the works.

Information to be produced by the parties

The same approach has led to the development of requirements that the contractor shall provide labour and material histograms for the whole of the contract period together with a resource schedule and method statement showing the techniques to be deployed over the course of the works, and a contract management chart ('organo-gram') showing who is to do what and to whom they are to report. This is frequently followed by a provision requiring the contractor to identify the dates by which he is likely to require the release of certain information.

Two other increasingly common provisions are those requiring specific testing programmes to be undertaken in relation to possible soil contamination and, particularly in civil engineering works, imposing a duty to produce a particular schedule for determining ground conditions with the object of limiting the possible impact either of clauses 11 or 12 of the ICE 6th or 7th edition or clauses 2 and 11 of the IMechE form. Appendix 1 sets out some sample provisions adapted from contract documents prepared over the past couple of years.[2.8]

Practical considerations

The advantages of this sort of régime are obvious: it allows the parties to have planned the campaign in such detail that they will know precisely what their obligations are and how they intend to fulfil them. The difficulties which are discussed below have a great deal less to do with the thinking behind these requirements than their practical application.

Critical path analysis

A common complaint of project managers and architects when faced with a programme which the contractor has produced under the sort of provision set out in Appendix 1 is that it will not work in practice. This frequently manifests itself in relation to the critical path network which is required under the model clause. The operation of such networks is considered in Chapter 7. However at this stage it is worth making a couple of observations;

(1) The expression 'critical path analysis' is not a magic formula – it describes a method of looking at the planning of the works which will enable the parties to determine how they can most efficiently be completed. Its particular attraction is that it enables the parties to identify which activities will be dictated by preceding or concurrent activities, and thus to target potentially crucial areas where delays would adversely affect other activities.

(2) Critical path networks can be produced in a number of forms. The Chartered Institute of Building produce an excellent introductory guide to their production and understanding. In recent years, a number of computer programmes have been devised for producing networks. They can, however, be

produced manually and in either case the key is not the complexity of the result but whether that result actually allows proper planning of the works. A great deal is made of the word 'logic' as applied to critical path analysis. The point is that if the network has been produced by some method which does not accord with a logical (which usually means 'possible') construction sequence, the whole process is of limited use.

This shows the strengths and weaknesses of the doctrine of providing the maximum amount of information either before or at the start of the works. While it will certainly allow the professional team to 'troubleshoot' and spot problems before they arise, it can also lead to the project starting in an atmosphere where the parties have already sown the seeds of mistrust over the practicalities of the programme. Whether or not this will be the case is very much up to the parties. The ramifications of this are considered later in this section.

Status of the programme

However, while this is a common criticism it is a little surprising that a much more pertinent point is seldom addressed. That is the status of the programme itself. There are two considerations here.

(1) Whether the programme has any contractual force i.e. do departures from it have any consequence? While the requirement to produce a programme comprises a step beyond clause 5.3 of JCT 98, this begs the question as to what reliance, if any, can be placed upon the programme.
(2) Whether or not the programme has some sort of contractual status, what is the effect of clause 2.2.1 of JCT 98 (or the equivalent clause 2.2 in WCD 98), the statement that nothing in the contract bills will override or modify the application or interpretation of the articles of agreement, the conditions or the Appendix, or the statement in clause 5.3.2 that nothing in the programme shall modify any obligation imposed by the contract documents?

The wording in the fist model clause in Appendix 1, while typical of its kind, does nothing to answer either question. It is all very well to say that the contractor shall produce a programme. Certain questions then arise which spring from the cases considered above:

- Is the contractor entitled to regard the programme as setting out a fixed itinerary for the works? Do events which lead to departures from it have a consequence under the variation and extension of time clauses?
- Is the contractor obliged to comply with the sequence which the programme contains, and if he does not can this give rise to a claim on the part of the employer, and if so what?
- Alternatively, is the programme merely a planning tool which exists for the convenience of the contractor and the reassurance of the employer and which has no other status of any sort?

Applying the ordinary rules of contract interpretation, namely that of giving the words their ordinary everyday meaning, the answer to the first two questions appears to be 'no'. Despite the complexity of the information to be provided, there is nothing to suggest that the programme can actually be relied upon in the same way as, say, the contract drawings. Hence, the application of clause 2.2.1 provides that the programme shall not modify the appendix provisions providing for the commencement and completion of the works. Thus, the parties' obligations are limited to the start and finish dates; the programme has no significance beyond its role as a guide to both parties.

The programme as a measuring instrument

That is not the end of the matter. In the event of disputes, two arguments will be run. Firstly that, notwithstanding the wording of the particular clause in the preliminaries, the parties themselves agreed that the programme would be the measuring instrument by which progress would be judged. The reality is that this is difficult to disprove and is the sort of argument which can be presented attractively to a court, adjudicator or arbitrator. If the dispute is simply one which is concerned with delay and its causes there is no very good reason why the employer should argue this point. It has the advantage of certainty and simplicity; the programme is a convenient yardstick.

Secondly it will be argued that in the majority of projects

employing relatively unamended standard forms it actually does not matter whether the programme is a 'contract document' or not. Under the sort of contractual scheme provided under the principal standard forms, the only relevant provision is that requiring the contractor to commence by a particular date and complete by a specified later date. The contractor will say that in order to fulfil that requirement he produced the programme. Provided that the programme was actually achievable, if departures from it lead to delays beyond the contract completion date then, almost by default, it will become the yardstick by which progress will be judged. The programme will not be a contract document but it will provide powerful evidence. Interestingly, the compensation events provision of NEC Second Edition is an exception to this. Clause 0.1(3) states that failure to provide something by the date set by the programme shall be a compensation event.

This latter point was neatly illustrated in *Glenlion Construction Ltd. v. The Guinness Trust* (1987) 39 BLR 89 at page 99. That case concerned a contract incorporating the conditions of JCT 63 which by clause 12(1) contained a provision very similar to clause 2.2.1 of JCT 98 and if anything went further by providing that nothing in the contract bills shall 'affect in any way whatsoever the application or interpretation' of the conditions. The bills nonetheless provided a requirement that the contractor should produce a programme in a specified form. Judge Fox-Andrews was asked whether this bill requirement was a contract provision and answered the question by stating that since the bills were a contract document, the particular clause was a contract provision.

While the particular question was important in the *Glenlion* case for reasons which will be discussed below it follows from the above that in most cases it does not really matter whether the programme is a contract document in the strict sense or whether it is simply a document which is produced to give effect to a bill provision of the sort found in *Glenlion* or the first model clause in Appendix 1.

Whether it will work

However, thus far we have assumed that the parties accept that the programme could be met. This will not always be the case where there is dispute as to whether the programme will actually work in practice.

The question of what will constitute a reasonable time has been considered above. This issue raises similar considerations. It will be appreciated that the workability or otherwise of a contractor's programme is a difficult problem. This is substantially because this is one of those issues where two people can hold diametrically opposed views but there is merit in both of their contentions. It is therefore precisely the sort of problem which is likely to require resolution by a court.

The task of finding a practical solution to the problem has not been helped by the two leading commentators. In the 10th edition of *Hudson's Building and Civil Engineering Contracts* Ian Duncan Wallace wrote that

> 'litigious contractors frequently supplied to architects or engineers at an early stage in the work highly optimistic programmes showing completion a considerable time ahead of the contract date. These documents are then used (a) to justify allegations that the information or possession has been supplied late by the architect or engineer and (b) to increase the alleged period of delay or to make a delay claim possible where the contract completion date has not in the event been extended.'

In the 11th edition it is conceded that this overlooks the practice which will be considered below of producing a programme incorporating a float to guard against unforeseen contingencies. The matter is not taken further.

In the supplement to the 4th edition of *Building Contracts* Donald Keating took a similar approach stating that

> 'a contractor does not prove a claim for delay in instructions merely by establishing non-compliance with requests for instructions or a schedule of dates for instructions he has served on the architect. But agreement with the architect with such a schedule or even acquiescence may it is submitted be relevant evidence as to the question of what is reasonable.'

In the 5th edition it is noted that this passage is quoted with approval in *Glenlion*.

To say the least, this is not ideal. The result of this is that the parties may well have been using a particular programme to gauge progress without a thought as to its legal significance. If disputes subsequently arise in relation to delays and a subsidiary argument occurs as to whether that programme marked the appropriate way to measure delays, the parties will be pitched into an argument which will be difficult and costly to resolve and which will inevitably delay the resolution of the principal dispute – that concerned with delays.

Avoiding such uncertainties

From a practical perspective, the issue is obvious – how do the parties guard against this sort of problem? The answer is fairly clear. If the author of the 'prelims' redrafted the programme requirement in the manner provided in model clauses 2 and 3 in Appendix 1, the uncertainty identified above would be eliminated. It will be seen that model clause 2 provides that notwithstanding the requirement that the contractor should submit a programme, it is not to be taken as having any contractual significance and, furthermore, the employer specifically reserves the right to issue instructions altering the sequencing of the works should he so desire, and that such instructions shall not of themselves entitle the contractor either to an extension of time or any other recompense. Model clause 3 approaches the matter from the opposite perspective and provides in essence that the programme shall be taken to set out the sequence of working and that instructions causing departures from it shall be taken to entitle the contractor to the appropriate extension of time.

That of course leaves us with the stipulation which appears in JCT 98 and WCD 98 that nothing in the bills or Employer's Requirements shall override or modify the application or interpretation of the contract conditions. Obviously, in *Glenlion* the possibility that the (apparently even more stringent) requirements of JCT 63 would operate to prevent a programme from being followed does not seem to have appealed to Judge Fox-Andrews, although it is unfortunate that he did not share his reasoning on this part of his decision. Perhaps the best view is that it is mistaken to regard any provision of this sort as overriding or modifying the contract conditions. Instead, what they do is merely to provide a machinery to enable the parties to operate the contract conditions. At any rate, it is difficult to imagine circumstances in which a court, arbitrator or adjudicator would be prepared to strike out a provision of this sort as offending against clause 2.2.1 or 2.2. The only possible exception to this is the situation which will be considered in Chapter 4 where the contractor produces a programme which puts forward dates for possession or completion which differ from those in the Appendix.

To summarise, it will be appreciated that the requirement that the contractor should provide a programme is often fraught with potential difficulties. The importance of these points goes not only to the planning of the works themselves, but is fundamentally important in subsequent delay claims and the manner in which such claims are approached. The fact that these problems originate in 'lawyers' arguments' as opposed to any practical consideration does not diminish their potential seriousness. Neither does the fact that the source of the problem in most instances will be an attempt on the part of the employer's professional team to exert greater control over the works. However, the real message is that these are arguments which can be avoided and, with slightly more precise drafting, future disputes can be nipped in the bud.

Further practical considerations

It is tempting to conclude that once the parties have sorted out the programming of the works, they have crossed the major obstacle to successful planning of the project. That may well be true to the extent that the programming of the works seems to give rise to a disproportionate number of disputes. However it may be sensible briefly to consider here and in subsequent sections some of the other considerations and sources of dissent connected with the various pieces of information required or provided as part of planning the works.

Just as criticisms are levelled at the contractor's programme, the requirement that he should produce labour and plant histograms or organisation schedules is frequently met by the complaint that they are either insufficient or impractical. This is a form of shorthand for suggesting that the contractor has not planned to devote the levels of resources which the employer's team consider should be allocated to the project and that, in order to deliver the works for a specific price, the contractor will need to cut corners.

It is possible that such criticism will identify a serious problem in the making and allow it to be averted. However it may do no more than highlight a dilemma from which there is no obvious escape. We have already briefly considered this problem in Section 2.1 in relation to the allocation of risk. It is at the planning stage that another manifestation of this particular problem appears – the contractor may be prepared to take on a greater degree of risk in a recession, but the key question is whether the resources which he can actually devote to the project will be sufficient.

The history of the construction industry in Great Britain during the first half of the 1990s is a catalogue of shrinking markets and falling profits. Even in the comparatively more prosperous period since then, the frequent complaint of many has been of growing workloads but static profits. In many sectors of the industry, projects priced below cost have become the norm. Hence it would be naive of a project manager to assume that the apparent under-resourcing of a job reflected either an oversight on the part of the contractor or an attempted shortcut, both of which could be remedied with additional resources allocated. It would be almost as unwise to say that, since the contractor had priced the job on that basis, the problem was his and nobody else's. The reality is that such projects are precisely those with a propensity to go wrong. Furthermore, and as we have already seen, common sense suggests that where the project has been tendered on the basis that it will actually lose money, the only way in which the contractor is going to break even on the job is through variations or claims.[2.9]

It would be defeatist to say that there is nothing to be done in this situation and that damage limitation is the only approach. At its simplest, the problem is that the works cannot be built for the price. However, it is sensible to acknowledge that in this sort of project the advantage of a system which requires the production of extensive detail prior to the works is that not only can the danger signs be spotted but the professional team can endeavour to minimise the problem.

As a rule of thumb, where the contractor has been guilty of 'buying work', the inadequate resources which can be dedicated to the job factored against the negative profit margins suggest that the contractor will probably achieve a worse result even than that suggested by the tender. On that basis if the project can be brought to completion without a major dispute that will constitute a minor success.

While of course it will depend upon the individual project, the following considerations may well be relevant:

(1) Recognising the likelihood of problems: these are precisely the projects where the professionals must be on their mettle. In essence, it is up to them to do everything to allow the contractor the best possibility of completing the job without conflict.

(2) The problem areas are likely to be those where the contractor will be required to expend substantial sums over a prolonged period and where there may be a lengthy gap before the contractor recoups this through interim certificates. The obvious examples are likely to be mechanical and electrical services.

(3) While the contractor will be looking to the variations account and to potential claims, his first priority should be the avoidance of allegations of failing regularly or diligently to progress the works with the attendant risk of liquidated and ascertained damages.

Finally it should be added that the comments made above about the application of clause 2.2.1 of JCT 98 (and its equivalents in the other standard forms) to programmes will apply equally in relation to the other types of information which may be required.

The proliferation of clauses such as those set out in Appendix 1 coincided with the recession. At a time when contractors, developers and professionals were working to ever-reduced margins, the pressures upon members of the industry to pre-plan projects appeared to be ever-increasing. Inevitably, as technology advances, the range and sophistication of planning tools will increase. However, the economic climate of the early 1990s also suggested an increased level of caution on the part of developers. Even in the context of an industry-wide recovery from the effects of the early 1990s depression, a guiding feature has been the wish to avoid the over-heated economy seen in the late 1980s. The provision of information can thus be seen as a type of insurance. The difficulty with this is that compliance with these requirements becomes an end in itself, with the result that parties appear to be going through the motions rather than seriously planning how to realise the particular project successfully. Needless to say, this is self-defeating.

Conclusions

The purpose of the provision of information and the use of the various planning tools is to enable the parties to put their respective contract obligations into effect. It can be reduced to a single question: 'How are we going to deliver this project on time and within budget?'

The simple truth is that the requirement to produce the sort of information considered above is designed to permit the parties properly to understand the project and everything it entails. That involves proper appreciation of the works themselves together with understanding the conditions both physical and contractual under which it is to be undertaken.

The planning of the project marks the first step towards success or failure. However, construction differs from sporting or military planning in that the game or battle is won on the site.[2.10] Nevertheless, there is an important intermediate stage between compliance with the information requirements contained in Appendix 1 and the commencement of operations on site. This is a phase in the construction process which does not really have a name but can best be described as 'design flow planning'. It is the process of determining the information which will be required sequentially during the project and the dates when it will be needed.

2.4   *Information–when and what*

There is no such thing as the perfect project. Seldom will the design of the project have reached a stage where all the necessary information is available to the contractor at the date for possession. The purpose of the clauses in Appendix 1 which are concerned with dates when information will be required is to ensure that the design team will be able to optimise the efficient production and flow of design information to enable the contractor to comply with the programme. That, of course is the case in traditional contracts. In design and build contracts the position is not significantly different, although the contractor will look to his own design team for the flow of necessary information. The point is reinforced by the checklist at Fig. 2.2.

The corollary to this is the provision of clause 25.4.6.in JCT 98 and its equivalent provisions in the other major standard forms. That identifies as one of the relevant events entitling the contractor to an extension of time the architect's failure to supply documents or information as required by clause 5.4.

The Information Release Schedule is intended to ensure that the contractor will receive information on the appropriate dates and that disputes and delays caused by late information can be avoided.

The difficulty is obvious. What information should be provided, when should it be provided, and when can it legitimately be requested? This is of course a question for which there is no 'legal' answer, it will depend upon the individual circumstances which prevail in each case. To the extent that the matter falls to be determined by the courts it could have occurred in two ways. The first is in relation to implied terms and these will be considered in the next section. The second and more general category is where there is a dispute either as to the adequacy of the contractor's application for information or in relation to whether the information was actually provided late.

In *London Borough of Merton* v. *Stanley Hugh Leach Ltd* (1985) 32 BLR 51 one of the many issues which fell to be determined by Mr Justice Vinelott was whether the contractor's programme presented at the commencement of the work could constitute an application for the purposes of clause 23(f) of JCT 63, which is in materially similar terms to the terms identified above. It was held that a programme which showed the dates when instructions and details would be required could be an application within the meaning of this clause, provided that, objectively determined, the dates of the requests were not unreasonably close to or distant from the due dates for the information.

This is helpful in that it puts paid to the idea that a contractor can lay the foundations for a claim by arriving at the pre-commencement site meeting armed with a schedule in which, in essence, he says, 'I want all of the information and I want it now'. That is a self-defeating approach. In some respects it indicates that the planning exercises considered above have served little purpose. The provisions of clause 31 of the NEC Second Edition make interesting reading, providing a comprehensive code for the timely provision of necessary information.

So what exactly is meant by 'in due time'. In *Percy Bilton* v. *GLC* [1982] 1 WLR 794 the expression was held to mean 'in a reasonable time' and not 'in order to avoid delay'. Plainly the two expressions will not always mean the same thing.

That does not provide an opinion on the really important issue – what information the parties can require. As stated above, this will be different in each instance. What is clear is that it is incumbent upon the contractor to attempt to determine what information really will be required and when. To take the example of excavation works on a street-lighting contract. The critical issue is likely to be when the excavation contractor will need to have details of ground conditions, in what form, and whether that is going to be required all at once or sequentially, depending upon the way in which the works are to be undertaken. Whether or not the contractor has actually been required to ascertain some or all of this information under the terms of the prelims is secondary to the fact that, in order for the works to be undertaken successfully, the contractor must enter the project with some sort of plan for dealing with these matters.

The key to successful planning and avoidance of subsequent delays therefore goes well beyond mere compliance with the preliminary requirements to produce information of particular types by certain dates. From the point of view of both the contractor and the employer, it is crucial to give thought to the precise manner in which the project is to be accomplished prior to the start of the works.

2.5 *Getting it right from the outset – contractual obligations*

Cautionary lessons

> The importance of planning the works properly is shown by the proliferation of bad claims caused by the parties doing a bad deal. In many instances this is caused by one or more of the parties not really understanding what the works will comprise. The examples of this are many and varied, although unsurprisingly, many involve the discovery by the parties of 'unforeseen' ground conditions, resulting in additional or delayed work. Typical is the case before the House of Lords in *Thorn* v. *London County Council* (1876) 1 App Cas 120. Contractors had agreed to demolish the old London Bridge and build its replacement. The design called for the foundations to be constructed using wrought iron caissons, the upper parts of

which were to be removed as the works progressed. These proved incapable of withstanding the tidal pressures and the foundations eventually had to be constructed at low water at greatly increased cost. The contractor contended that the specification was to be subject to the implied term that it could be constructed relatively inexpensively in accordance with the design. The House of Lords gave this argument short shrift, and Lord Hatherley's comments are typical:

> 'If there can be found any warranty in such a contract as this....it would scarcely be possible for any person whatever to enter upon any new work of any description'.

The difficulty which confronts many unwary contractors is that the hazard only manifests itself after they are committed to carrying out the work, and in many cases after they have started work. In *Bottoms* v. *York Corporation* (1892) 2 HBC 208 the contractor, having agreed to carry out sewerage works on land adjoining a river without having taken the precaution of sinking any boreholes, discovered that the soil conditions rendered his original construction method impossible. As with the decision in *Thorn*, the court refused to allow him to claim the additional cost to which he had been put, occasioned by a different method being needed for the works which took longer. The principle is simple; at its bluntest, the fact that a party has undertaken a bad bargain is his own and nobody else's fault.

It also provides a cautionary lesson. Contrary to the view which is widely held, the mere fact that information is supplied to the contractor at the time of tender does not amount to a warranty that it is accurate or that the works themselves are feasible. Indeed, the approach taken by the courts has been to resist implying any such warranty except where there is clear evidence permitting them to find the sort of warranty, contended for and rejected in both *Thorn* and *Bottoms*. Exceptionally in *Bacal Construction* v. *Northampton District Council* (1975) 8 BLR 88 it was found that there had actually been an implied warranty by the employer that the ground conditions would accord with the hypothesis upon which the contractor had tendered, but this depended upon the particular facts.

Unless the contractor can bring himself within this exception, he is in difficulties. It is also worth noting that in this instance the contractor will not be able to rely upon any of the doctrines of misrepresentation, mistake or frustration.

(1)   There cannot be a misrepresentation because there is no warranty that the statements supplied at the time of tender were actually true. Hence, if the contractor relied upon them that was his lookout.

(2)   If the contractor either believes the information supplied to him or fails to have independent checks carried out this is not a mistake in legal sense.

(3)   It is clear from *Thorn* that frustration can only apply where the works have been so changed that the original contract ceases to have any application. It is plain from the relatively recent decision in *McAlpine Humberoak* v. *MacDermott International* (1992) 58 BLR 1 that in the context of a construction contract frustration is only going to occur in the most exceptional circumstances.

Hence one of the parties has obtained something very different from his expectations. There are no statistics to demonstrate the split between instances where this has occurred because of a difference as to the meaning of the contract conditions or because of a difference as to the exact nature of the works, or a combination of the two. However, it is reasonable to assume that a very high proportion of such contracts occurred either because one or both parties simply assumed that a certain state of affairs prevailed, or they did not think about it at all, or they decided to take a commercial view – or gamble.

It is also sensible to remember that the assumption that a certain situation exists may not accord with the rules governing the construction of documents. As has been stated above, a contract is construed by reference to the ordinary everyday meaning of the words which are used. While there are limited circumstances in which the court will pay attention to the surrounding 'matrix' of facts in which the relevant words are used, that does not mean that any regard will be had to the 'reasonable expectations' of the parties, or indeed to their intentions, or even what had occurred on previous contracts between these or other parties.

While a very high proportion of these problems could be avoided by a certain amount of caution on the part of the contracting parties at the time of the agreement, hindsight is a wonderful thing. This leads to a certain number of attempts by parties to rewrite, or at any rate, to redefine the bargain. These tend to take three forms:

- writing further obligations into the contract by way of implied terms;
- redefining possibly ambiguous terms by what is called the *contra proferentem* rule; or
- calling into effect the provisions of the Unfair Contract Terms Act 1977.

Implied terms

Frequent attempts have been made by the parties to construction contracts to fill in the gaps in their contract by way of implied terms. Interestingly, a great many of these have occurred in circumstances where events have caused the works to take longer or become more difficult. Such attempts have enjoyed varied degrees of success in the courts.

Implication of terms occurs in two ways – by operation of law such as under sections 12 to 15 of the Sale of Goods Act 1979 and sections 13 to 15 of the Supply of Goods and Services Act 1982, or by necessary implication.

In the context of construction contracts necessary implication is best described in the 5th edition of *Keating on Building Contracts*:

'The test of implication is therefore necessity – "such obligation should be read into the contract as the nature of the contract itself implicitly requires, no more no less".[2.11] The term sought to be implied must be one without which the whole transaction would become "inefficacious, futile and absurd".[2.12] The implication may be necessary because "language is imperfect and there may be, as it were obvious interstices in what is expressed which have to be filled up".[2.13] What the court does, if so persuaded, is in effect "to rectify a particular – often very detailed – contract by inserting in it a term which the parties have not expressed".[2.14] A term sought to be implied under this heading has to be what the parties must have intended and therefore is not implied if it provides only one of several possible solutions to the matter in question.'

Necessary implication is expressed in a number of different ways; it is frequently said that a particular term is to be implied either to give business efficacy to the contract or to give effect to the common intention of the parties. Both of these expressions mean the same thing – that the term is necessary in order to make the contract work. However, the very nature of implied terms means that the terms to be implied into a particular contract will depend upon the precise circumstances and there is no general set of terms which will be implied as a matter of course. As Lord Justice Lawton said in *Martin Grant & Co Ltd* v. *Sir Lindsay Parkinson & Co Ltd* (1984) 29 BLR 31 at page 41:

> 'building contracts have regularly been tried in this court all my professional lifetime and long before. If there were a general rule ... it is surprising that it has never been recognised before.'

The reason for this is clearly set out in the speech of Lord Simon in *Luxor (Eastbourne) Ltd* v. *Cooper* [1941] AC 108:

> 'There is, I think, considerable difficulty and no little danger, in trying to formulate general propositions on such a subject, for contracts with commission agents do not follow a single pattern and the primary necessity in each instance is to ascertain with precision what are the express terms of the particular contract under discussion, and then to consider whether these express terms necessitate the addition by implication of other terms.'

This is equally applicable to construction contracts. The express terms will vary according to the nature of the works and hence there can be no general rule.

It is sometimes argued that implied terms mark a way in which contracts can be re-written. This will not be permitted where the effect of the implied term which is contended for will be to contradict the express wording of the document. Thus, while it is possible to identify a number of terms which have commonly been implied, all of those must be regarded as subject to the express provisions of the contract.

In *Luxor* Lord Simon said that, generally speaking where, 'B is employed by A to do a piece of work which requires A's co-operation ... it is implied that the necessary co-operation will be

39

forthcoming'. This has also been expressed as the employer impliedly promising to do everything necessary on his part to bring
about completion of the contract. The exact manner in which that
will be required will vary according to the circumstances.

- in *Hounslow LBC* v. *Twickenham Garden Developments* [1971] Ch233
  this included a warranty not to revoke the contractor's licence to
  occupy the site;
- in *Panamena* v. *Leyland* [1947] AC 428 it was held that it will
  ordinarily be the responsibility of the Employer to require the
  certifier properly to perform his duties;
- in *Glenlion* v. *Guinness Trust* (see above) instructions and details
  should be provided at such time and in such manner as not to
  hinder or prevent the contractor from carrying out his duties
  under the contract.

However, it is easy to fall into the trap of over-simplifying the
implication of terms, particularly where they affect timing obligations. The last example given above demonstrates this. In *Glenlion*
the contractor argued that he should be entitled to carry out the
works in accordance with a programme. He went on to argue that,
since the programme showed an earlier completion date than the
contract completion date, he was not only entitled but also obliged
to complete by this date and that acts of prevention by the employer
producing a later completion date (albeit one which was still prior
to the contract completion date) would, in effect, entitle him to claim
extensions of time. Thus he contended that the contract should be
read as being subject to an implied term varying the completion
date. This argument was rejected by Judge Fox-Andrews who held
that the term argued for was not so self-evident that it went without
saying, therefore it could not be implied 'of necessity'.

In reaching this conclusion, the judge held that the contract as
drafted – that is without the implied term – was efficacious and
produced the desired result.

Attempting to import implied terms which may be appropriate to
one contract into another is a very real problem where the parties
are dealing on a non-standard contract, particularly one which may
have been created with a particular project in mind. This was
considered in the *Martin Grant* case (see above), where the sub-
contractors contended that the sub-contract was to be read as being

subject to the implied term that the contractor would make sufficient work available to the sub-contractors to enable them to
maintain reasonable progress and to execute their work in an efficient and economic manner and that the contractor would not
hinder or prevent the sub-contractors in the execution of the works.
Mr Justice Lawton followed the decision at first instance of Sir
William Stabb and held that the sub-contract had no room in it for
the implication of this term which was not consistent with the
express wording of clause 3 which made the sub-contract sub-
servient to the Main Contract Works. The interesting feature of this
case is that, in the absence of this express clause, the implied term
was one which might have been incorporated and had been successfully argued elsewhere.

The need for caution in seeking to imply terms is clear. As Lord
Pearson said in *Trollope & Colls* v. *North West Metropolitan Regional
Hospital Board* [1973] 1 WLR 601

> 'The court will not even improve the contract which the parties
> have made for themselves, however desirable that may be. The
> court's function is to interpret and apply the contract which the
> parties have made for themselves. If the express terms are per-
> fectly clear and free from ambiguity, there is no choice to be made
> between different possible meanings: the clear terms must be
> applied even if the court thinks some other term would have been
> more suitable.'

In other words, the court will not come to the rescue of parties
who have made a mess of sorting out their contractual affairs by
way of rewriting the contract by means of a series of convenient
implied terms. As we have seen above, there are no 'usual' terms,
nor is it the case that the court will produce a construction of the
contract which is 'reasonable' in circumstances where some other
meaning is clear. This point is worth bearing in mind when considering the second limb of this section, that of attempts by the
parties to rewrite, redefine or strike out parts of the contract.

The *contra proferentem* rule

This is another expression which is frequently used inaccurately. It
covers the situation where there is an ambiguity in the express words

of a document and all other attempts to resolve the uncertainty have failed. In those circumstances, the document will be construed against the person who proffers the document. It will give effect to the meaning which is favourable to the other party. This should almost invariably mean that the document will have been drafted by one of the parties and imposed upon the other. The thinking behind the rule is not difficult to work out; if a party has imposed his terms on another he must be taken to understand what those terms are supposed to mean and the court will be unsympathetic towards attempts by him to resolve ambiguities in his favour.

However, a certain amount of judicial confusion seems to have occurred. The result is to raise the possibility that the rule may be applicable to standard form contracts. Keating[2.15] points out that the expression should not be applied in these circumstances where, after all, the contract has been drafted by a body drawn from a cross-section of the industry. He observes correctly that the expression has been used in a wider sense to deal with ambiguities in the liquidated damages and extension of time provisions in standard forms. This is little help to the contractor or employer engaged in a dispute as to the meaning of a particular provision – does this particular rule apply or not?

The answer appears to be that it does not. Writing about its application to liquidated and ascertained damages, Brian Eggleston noted that the courts have shown a noted lack of enthusiasm for the rule and said that in his view it was odd that the rule should apply to standard form contracts.[2.16] On closer examination, it can be seen that they do not really apply the rule.

(1) The starting point is *Peak Construction (Liverpool)* v. *McKinney Foundations* (1970) 1 BLR 114. In that case, Lord Justice Salmon said that:

> 'The liquidated damages and extension clauses in printed forms of contract must be construed strictly *contra proferentem*.'

However, that case concerned a clause in a contract which was not a standard form but rather was the employer's own form, albeit that it was substantially derived from JCT 63. The liquidated damages clause was indeed construed strictly against the employer.

(2) This case was cited in *Rapid Building Group Ltd* v. *Ealing Family Housing Association Ltd* (1984) 29 BLR 5. Lord Justice Lloyd noted that it was common ground that in the light of *Peak Construction* that no liquidated damages could be claimed by the employer where the employer was the cause of the delay. This, of course, does not depend upon the *contra proferentem* rule at all. Lord Justice Stephenson, the other member of the Court of Appeal, approached the matter similarly and accordingly we can conclude that the reference to the *contra proferentem* rule is really no more than a red herring.

(3) Much the same can be said in relation to *Brammall & Ogden* v. *Sheffield City Council* (1983) 29 BLR 73. Although counsel for the applicants referred to the rule in relation to the operation of liquidated damages provisions in a standard form, this does not actually seem to have played any part in the decision of Judge Hawser, who made a clear finding on the facts of the case that the clause in question was to be construed in a particular way which prevented the employer from claiming liquidated damages. To have done otherwise and to have found any other meaning would have required that the contract be 'tinkered with' in a way which the parties could have done when they entered into the agreement but did not.

Accordingly, the rule is best seen as something which has application only in limited circumstances where the contract is not a standard form. This will be the case where the parties have attempted to rewrite the conventional wisdom regarding the entitlement to extensions of time and in so doing have created an ambiguity. That was the case in *Rosehaugh Stanhope* v. *Redpath Dorman Long* (1990) 26 CLR 80. The dispute concerned a tailor-made construction management contract which sought to limit the defendant's right of set-off and to extensions of time, while permitting the construction manager to ascertain and deduct sums arising from a failure to complete by the contract completion date. Lord Justice Bingham held that the provisions in question were ambiguous and hence, since the document emanated from the employer, he would adopt the construction less favourable to them.

The rule will be of no assistance where the parties are simply debating the construction of a standard form. This is obviously desirable – it avoids a patent absurdity, that if neither party has actually proffered the document, it would be perverse to construe it against either one of them.

Hence, while provisions regarding liquidated damages and extensions of time are often attacked, there is no general rule which provides that they, more than any other clause should be susceptible to attack on this basis. This is demonstrated by *Kitsons Sheet Metal Ltd* v. *Matthew Hall Ltd* (1989) 47 BLR 82 (see below). The rule was cited in argument to the judge. However, it played no direct part in the reasoning, nor really in the argument where both parties submitted that, on their reading of the contract, its meaning was clear and the judge was faced with a straight choice between two conflicting interpretations. It was not a case where there was a suggestion that there was an uncertainty on the face of the documents.

Exclusion clauses

The alternative to asking the court to redefine a particular clause is to ask it to strike out the clause altogether. From the point of view of the contracting party who has discovered that he has entered into an undesirable bargain, this is obviously a desirable prospect. Predictably, except in very limited circumstances, just as the courts have always taken the view that they will not rewrite a bargain, still less will they be inclined to delete sections at the request of a disgruntled party.

The only one of the limited exceptions to this general rule which is relevant to delay claims is the exclusion clause. It is not difficult to imagine a situation where the parties have drafted the extension of time clause in a contract in such a way as to confine or remove the entitlement to an extension of time. More commonly, one of the parties will have sought to effect some limitation on the right to claim damages arising from a breach of contract, frequently seeking to exclude, for example, damages arising from delays occasioned by the breach. This is frequently the case in supply contracts where a party will seek to escape from the consequences of late delivery by means of an exclusion or limitation of damage clause.

The general rule is that commercial entities contracting together are bound by the bargains they make. The leading case is *Photo Productions* v. *Securicor Transport* [1980] AC 827 where the House of

Lords held that, despite the circumstances of the case (the plaintiffs' factory was destroyed by a fire deliberately started by the night watchman, the defendants' employee), it was impossible to find any reason why an exclusion clause protecting the defendants from any liability except for the negligence of their employees should not mean exactly what it said on its face, and that therefore Securicor should be entitled to rely upon it. Lord Salmon summed up the position neatly:

> 'Any persons capable of making a contract are free to enter into any contract they may choose; and providing the contract is not illegal or avoidable it is binding upon them ... In the end everything depends upon the true construction of the clause in question.'

The reasoning of the House of Lords is clear; where a contract is entered by businessmen who are capable of looking after themselves and apportioning risks, the court should be reluctant to turn clear words on their heads in order to undo the bargain which they have made.

The one clear exception to this is found in the Unfair Contract Terms Act 1977. The application of this statute is widely misunderstood, not least because, like many Acts of Parliament, it is imprecisely drafted and some of the confusion which it is designed to alleviate is actually increased. For present purposes, the Act affects situations where one party either deals on the standard terms of business of the other or deals as a consumer (i.e. not in the course of business) and provides that where a term seeks either to limit the liability of the other for breach of a contract term or permits different performance from that which would otherwise be expected or no performance at all that term is only valid if it

> 'shall have been a fair and reasonable one to be included having regard to the circumstances which were, or ought reasonably to have been, known to or in the contemplation of the parties when the contract was made.'

The most common application of this situation to construction contracts will once again come in relation to liquidated and ascertained damages provisions. It is also easy to envisage it applying to

situations where one party seeks to limit or redefine the circumstances in which one party will be entitled to claim extensions of
time. This is particularly the case in the standard forms of domestic
sub-contracts devised by some of the major contractors and
imposed upon their sub-contractors. As is stated above it is also
frequently encountered in relation to supply contracts where the
supplier seeks to excuse the consequences of late or non-existent
delivery.

However, before the act can be invoked there are three hurdles to
overcome, the first of which has two limbs.

(1)   That the party upon whom the terms are imposed deals as a
        consumer. In construction contracts this produces the poten-
        tially anomalous result that where an employer who is not
        primarily engaged in the construction industry procures a
        building, he may be able to rely upon the Act, whereas, a
        property developer would not. Whether therefore, a Uni-
        versity or Health Authority would be protected by the Act is
        unclear. The same question could be applied to whether a
        Local Authority housing or works department would be out-
        side the Act, while perhaps its amenities department would
        not. The question has never been tested and for this reason
        should sound a warning bell.
(2)   The alternative is for the party seeking to attack a clause to
        show that he dealt on the other's standard conditions. This
        prompts the now familiar question of whether the use of a
        standard form contract can fall within this definition. Again,
        this has scope to throw out some odd results – does one party
        habitually trade on those terms, or have the material provi-
        sions been tinkered with in some way. The result may depend
        upon little more than a series of accidental factors.

There is disagreement among the commentators as to the exact
extent of the Act's application. *Keating* identifies the problem
without offering a view as to the answer. *Emden*, on the other hand,
suggests that there is no reason why it should not apply in some of
the situations identified above without providing a test to determine which cases will be caught by the Act. Brian Eggleston, in the
context of liquidated damages, provides limited support for this
latter view, while also acknowledging the observations of Lord

Justice Pearson in *Tersons Ltd.* v. *Stevenage Development Corporation* (1963) 5 BLR 54 (a case decided some 14 years before the Act and therefore of only illustrative value) where it was held that in relation to a standard form of contract it was hard to say that it was one side or the other's document. Put simply, the position is most unsatisfactory.

However, and assuming that this particular problem can be negotiated, the other two hurdles remain. The first is the requirement that the clause actually limits or excludes liability. Common sense suggests that arguments will seldom arise – it is difficult to imagine a party contending that a particular clause does not limit or exclude liability. A possible exception might be where an attempt is made to replace liability for a breach causing damages with an obligation to insure against the consequences of such breach. It is not known exactly how this argument would fare.

The most common and most argued over application of the Act is in relation to the test of reasonableness. The problem is that the Act contemplates that each situation will need to be judged on its own facts. Thus the cases provide illustrations and little else. One of the more helpful is *Rees Hough Ltd* v. *Redland Reinforced Plastics Ltd* (1984) 27 BLR 136. In that case the supplier dealt upon a set of terms which he had produced himself and which excluded all liability for losses sustained as a result of defects in the goods. The goods, reinforced concrete pipes, were defective and were the cause of the contractors abandoning the planned pipe jacking in favour of another method of construction. Judge Newey painstakingly analysed all of the factors both for and against the finding of reasonableness, and eventually came down firmly against the suppliers.

Obviously, the application of the Act to construction contracts will not be particularly common. Indeed, it may be a hit or miss affair to show that it is relevant to particular facts. However, it is plain that in a limited range of contracts it will be a material consideration and in these contracts it is unhelpful but unfortunately accurate to say that the wording of the Act prevents any degree of certainty. The highest that it can be put is that the courts will tend to look askance at a provision which seeks to exclude liability altogether. They will be more sympathetic to the sort of provision which limits the consequences of a breach to some readily defined sum which does at least allow the parties to argue that they had considered the effects of a breach of contract and had attempted to apply a sensible formula to determining them.

Express terms

It is worth briefly considering two cases which provide further illustration of the snares which can await those who leave any part of the timing of the works to chance. Neither involves implied terms because in each, the terms of the contract were clear, at least to the court, if not to the parties.

The first is the 1980 decision of the Court of Appeal in *M Harrison & Co (Leeds) Ltd* v. *Leeds City Council* (1980) 14 BLR 118. In that case, the architect, one JGL Poulson had obtained quotations from a proposed nominated steelwork sub-contractor which included a condition relating to the ground conditions which, as he was aware, conflicted with the contractor's intended programme for the works and could not have been implemented. Nonetheless he issued an instruction to place a nominated sub-contract with that sub-contractor including the condition. Subsequently, to overcome this problem, the contractor and sub-contractor agreed a revised programme for the steelworks and the sub-contractor sought and was paid an additional sum to reflect the additional time which had been taken to execute their work. The court held that the effect of this instruction was to operate as an instruction for the purposes of clause 21(2) of JCT 63 postponing the steelworks which in the event had been executed at a different and later time from that originally envisaged by either the contractor or sub-contractor. As the editors of *Building Law Reports* observe, construing the postponement instruction by reference to the contractor's programme has extensive implications. Not the least of these is as a warning to parties who seek to overcome a problem which has arisen during the negotiating process by way of a fiction. The rules of construction of documents provide that if this is what the contract says, then the fact that some other result may have been in the minds of the parties is neither here nor there.

The second case is *Kitsons Sheet Metal Limited* v. *Matthew Hall Mechanical and Electrical Engineers Limited* (1989) 47 BLR 82, a decision of Judge John Newey. The case concerned the construction of Terminal 4 at Heathrow Airport. The defendants were mechanical services sub-contractors and the plaintiffs were sub-sub-contractors engaged to carry out insulation work. The sub-sub-contract was in a form of the defendant's own devising. While it contained clauses entitling the plaintiff to claim extensions of time and loss and expense, these were in an extremely restricted form. The plaintiffs were, however, entitled to claim that variations would include not only variations in the quality or quantity of the work but also in the 'manner or sequence' of working. Additionally it was provided that the plaintiffs would work in accordance with the dictates or instructions of the defendant as they were from time to time issued, so as to complete by a specified date. (This last point is considered further in Section 3.4)

The plaintiffs' claim was in some respects akin to a disruption claim in that they alleged not only that the actions of the defendant had caused their works to take longer but that they had been rendered more complex by reason of what they said were alterations in the manner or sequence of working. These therefore fell to be dealt with as variations and they were to be computed by reference to the plaintiffs' programme. The defendants contended that in fact the wording requiring that the plaintiffs should work in accordance with their instructions meant that the plaintiffs were in essence obliged to work at the 'beck and call' of the defendants.

The judge found that:

> 'in accordance with the dictates (or instructions) of the defendant's management team are plain words, which admit of only one meaning, which can alternatively be expressed as "do what the management team orders".'

He went on to find that the circumstances of the project meant that detailed forward planning was not feasible and that hence it was imperative to adopt the utmost flexibility. Programmes were therefore useful but would change to suit varying circumstances. They could not have contractual effect. Hence departure from the programme could not properly amount to a variation.

This case provides an object lesson to those who enter into home-made contracts, particularly ones which only peripherally suit the works to which they are supposed to relate. It also provides a warning in the wake of the *Wells*, *Glenlion* and *Martin Grant* cases discussed above. Judge Newey applied the wording of the contract strictly. The consequence of this was that the programmes which had plainly occupied great quantities of the parties' energies both to produce and to implement were held to be no more than aids to the planning of the works. While the precise facts in this case are unusual, the provision in a contract requiring that one party should work at the beck and call of the other is potentially commonplace, particularly in management contracting or construction management contracts. Where the parties agree, in effect, that the programme shall not have contractual status, later attempts to argue to the contrary will not avail them. Perhaps the final, although most obvious lesson is that in producing a contract both parties should take immense care to ensure that their obligations are clear and unambiguous.

2.6   The employer's perspective

Most of this chapter has been approached from the viewpoint of the contractor considering the planning of a project and the pitfalls which he should avoid. Exactly the same considerations arise for the employer and indeed the seminal question for the contractor (How do we get the job completed on time and within budget?) will apply equally to the employer. Without wishing to take a needless pot-shot at Latham, the fact of the matter is that the parties to a contract do naturally start from the same objectives. The contractor or developer who starts from the perspective that his objective is to score points at the expense of the other is a relatively rare creature. The contractor or employer who knows exactly what the rules are and plays strictly in accordance with them is a less unusual phenomenon, but one which is perhaps still less common than should be the case. Condemnation of this approach, and by implication of all things 'contractual' (a term which has acquired the status of an insult in some quarters), seems difficult to justify.

Instead, the great enemy facing all sides of the industry is that of parties of all disciplines entering bargains which are inadequately thought out and then badly planned. It is obviously easy for lawyers to put forward a counsel of perfection. This attracts the criticism that such advice is easily offered from the safety of an ivory tower. The view that lawyers play too great a role in construction contracts would be easier to justify if the part played by them were not so heavily dictated by the desire of parties to unravel bad bargains.

Hence, for the employer, the planning and contractual checks which are dealt with above are aimed at an objective substantially identical to that confronting the contractor and in the final analysis each is concerned with ensuring that the project meets his original aspirations, both constructional and financial. In the following chapters consideration will be given to how the parties and the professional team can put these into operation and thus avoid delay.

CHAPTER THREE
DURING THE WORKS

3.1   Site organisation and reporting systems

The previous chapter dealt with the considerations which need to be addressed prior to the start of the works. There is no doubt that these present a bewilderingly large number of things for the parties to think about and the allegation could be made that this detracts from the essential business of building or civil engineering. Equally, without proper consideration of these matters, chaos will result. So, what does this have to do with an analysis of delay claims? The answer rather leaps from the page. The projects which involve substantial delays and where either party still makes a profit are the exception rather than the rule. In the previous chapter, one of the points considered was the problems associated with projects tendered at prices below break-even point. As a rule of thumb, where those projects can be brought back to a 'nil-nil' position where both parties break even, that marks a substantial achievement on the part of all concerned. In other words, delay claims are not to be sought out. If possible, they should be avoided. Part of that task is accomplished before any work is done on site.

The remainder is substantially achieved during the works themselves. Again, the central issue is the balancing act between time, quality and budget.

Relationships

This balance will be greatly affected by the relationships between the individuals most closely connected with the execution of the works themselves. In this respect there is no doubt that Sir Michael Latham has identified an important feature of the construction process. In paragraph 6.43 of his report he discusses the possibility of 'partnering' relationships between contractors and employers as a way to achieve a greater degree of harmony. The point can actually be taken further. Few people involved with the construction industry will dispute that if the parties' respective site teams fail to get on together the risk of the project going badly wrong will increase.

In this respect the relations between the senior management of the parties are far less important than the on-site relationships which are forged (or not) during the works themselves. This is a combination of the parties making the most of planning the works, together with all parties adopting an approach which is dedicated towards achieving the successful completion of the works.

The former demonstrates that the matters covered in the previous chapter are not merely window dressing. In the all too common scenario where the parties have dedicated insufficient energy towards planning the works, the chances are that they will set about the project with different understandings of their respective obligations. This will automatically place strain upon the site-based personnel and inhibit the prospects of them actually forming the intention to work together rather than turning the works into a battle of wits and nerves.

The latter is a much more difficult business. Both Latham and the authors of the NEC have dealt at length with the requirement for good faith between the parties in their business dealings. However, common sense suggests that this cannot be imposed, either by way of a contractual requirement or by government edict. Obviously, this is not a point which can be proved by statistics, but it is interesting to look at a couple of particular examples. Those are the Broadgate and Beaufort House developments in the City of London. Both were carried out under construction management contracts and both contained provisions whereby if a trade contractor failed to complete the works by the prescribed completion dates the employer would be entitled to sums to be estimated by the construction manager. In each project works package contractors failed to complete by the due date and the respective construction managers estimated that massive sums would be payable as a result. The employers commenced proceedings and sought summary judgment under Order 14 of the Rules of the Supreme Court (which then applied), and the contractors issued their own proceedings seeking declarations that they were entitled to extensions of time.

Both disputes (*Rosehaugh Stanhope (Broadgate phase 6 and phase 7) Ltd* v. *Redpath Dorman Long Ltd* and *Beaufort House Developments Ltd* v. *Zimmcor* (1990) 50 BLR 69 and 91) were heard at much the same time, before Judge Bowsher and later in the Court of Appeal. The latter held unanimously that, in order to operate this kind of set-off machinery, the employer had to show a breach on the part of the trade contractor. However, quite apart from the legal ramifications of the decisions, there is a wider issue. In both cases it is apparent that the parties adopted an unequivocally confrontational approach to the execution of the works. A trade contract which almost encourages the employer to react to delays by preparing claims which are not linked necessarily to any breach of contract will automatically foster disputes and this will be compounded where, as here, it is clear that the construction manager will make full use of such powers. It almost goes without saying that in this situation it is unlikely that the trade contractors concerned would have agreed to such a term had they considered its possible consequences.

Much has been written about the need to avoid conflict by way of enhanced interpersonal skills during the course of the works. This principle lies at the heart of the Latham Report. It follows from the previous paragraph that, in some instances, the cause of strife will be inherent in the contract itself. However, it is equally clear that, irrespective of the contract form, provided that both parties understand what they are supposed to be doing it will take more than an onerous contract to trigger a dispute.

It is almost a truism to say that a delay claim usually results because one of the parties fails to perform. One of the principal causes of this is a failure properly to plan the works or to get to grips with the obligations which the contract imposes. If this cannot be blamed, the likelihood is that the fault will be found 'at the sharp end', in the execution of the works themselves. Frequently this can be put down to a breakdown in one of the key relationships on site. There is an almost irresistible temptation upon contract managers to blame architects, employers' representatives and resident engineers for any problem, and this is almost always reciprocated. There will undoubtedly be instances where the individuals concerned simply do not and never will get on. However, this is seldom the whole story. More often than not the deterioration in relations is simply the manifestation of a much deeper problem in the works.

Planning to avoid conflict

This begs the question – even if a serious problem has occurred, is it inevitable that the parties will automatically draw up the battle lines? Alternatively, is it more desirable for them to resolve to make the best of things, endeavour to minimise the problem and work together to achieve the best result which circumstances permit? As a very broad rule of thumb, the risk of 'going legal' is implicit in the former approach, while in the latter, even if significant delays occur, the prevailing goodwill may serve to facilitate compromise. We are obviously dealing with hypothetical considerations. Seldom will any member of either party's team consciously resolve to pull the wagons into a circle and prepare for war. However, it is probable that every person who is engaged in advising parties to construction disputes will have seen a number of cases where the parties to a dispute seem to have embarked upon a headlong charge towards litigation. Common to such contracts is a feeling that whatever is said or done will merely serve to make the matter worse.

Although there are contracts where the extent of the personal animosity suggests that the answer lies either in counselling or duelling pistols, in most instances the problem is manageable. Obviously each problem of this sort is peculiar to its own facts. However, part of the solution lies in the increasing care taken by certain contractors, sub-contractors and employers to endeavour to put together site teams who will not only complement each other but will also react positively to the other parties in the construction process. This is an important trend and marks the realisation that conflict either causes delay or makes it worse; in such an atmosphere disputes lead to litigation or arbitration, and these are not the answer. At best they are damage limitation exercises, and at worst they take up vast amounts of costs, time and effort.

Unlike most of manufacturing industry, contracting generally involves a high degree of input by the customer. Except for so-called 'turnkey' contracts, the employer and his team can and do shape the finished product while it is under construction. As a result, the scope for conflict is greater than in most other industries where the purchaser is only really concerned to buy the end product. Save for extremely luxurious models, cars are available in a limited range of 'off the peg' types and there is little scope for the buyer to affect what he gets.

Key issues

The reason for this, as stated in Chapter 1, is that construction is concerned with the production of prototypes – hence the involvement of the ultimate owner of the prototype. However it is unsurprising that the scope for disputes is at its greatest at the point of day to day contact between the employer and the contractor or between the contractor and sub-contractor. In many cases, the problem is one caused by a failure to communicate effectively. This applies not only between the parties but also between the individual members of each party's team.

In this environment it is a toss up whether delays or defects, or a combination of the two, will result. It rather depends upon whether the incompetence of the contractor outweighs that of the employer's team. To the extent that such projects result in litigation, this is usually concerned with only the tip of a much larger iceberg as lawyers and claims consultants are generally concerned in this sort of situation to fit administrative disorder into some kind of legal or contractual framework. The reports of such cases seldom tell the whole story and in any event tend to address the legal arguments which will grow out of the events during the contract but rarely reflect exactly what went on. The decision in *West Faulkner Associates* v. *London Borough of Newham* (1994) 44 CLR 144 illustrates this. It is clear that the works descended from disorder into chaos and then into utter stagnation. Why this was so is not really dealt with in the law report which is simply concerned with the meaning of the expression 'regularly and diligently' in clause 27 of JCT 80.

However, it is not too hard to guess in many instances what has gone wrong. The problems variously result from:

- A failure on the part of the contractor to assemble a site team who can get on with one another in order to integrate their various functions. This is frequently compounded by a lack of continuity in the site team.
- The lack of adequate means of communication between site and head office so that different messages get back to head office to those which will be apparent to those on site.
- Clashes between the contractor's site team and that of the architect, engineer or employer.
- Breakdowns in communication between members of the

employer's team, resulting, for example in failure by the professional quantity surveyor (PQS) to treat variations in the manner envisaged by the architect when he instructed them.
- The inability on the part of either the employer, the contractor or both properly to organise works contractors, suppliers and sub-contractors.

The architect's role

Among the most frequently encountered disputes are those between the contractor and the supervising officer. Again, the reported cases are almost invariably concerned to consider the legal issues which arise from breakdowns in this relationship and largely fail to provide much in the way of explanation for this. In some cases it is possible to read between the lines. In *Lubenham Fidelities* v. *South Pembrokeshire District Council* (1986) 33 BLR 39, the court famously referred to the architect doing his 'incompetent best' and the attentive reader of the law report will derive the sense that the supervising officer had managed to alienate everyone with whom he had dealt not only by failing to understand his own obligations but also in erring at every stage in relation to the administration of the building contract.

In *West Faulkner* v. *London Borough of Newham* (referred to above), while the law report does not deal in any detail with the cause of the undoubted problems in the works, it does appear reasonable to conclude that the architect had, without taking professional advice, formed the view that the meaning of the determination clause in the contract was one which was at odds with the proper interpretation of the words used. More importantly, from a practical viewpoint, the architect seems to have failed utterly to carry out his super-visory function in a way which would either compel or cajole the contractor into improving his performance.

In *Design Liability in the Construction Industry* David Cornes provides a comprehensive analysis of the duties of the architect to supervise and inspect the works. He also refers approvingly to *Achieving Quality on Building Sites* published by the National Economic Development Office. As he says, this report should be compulsory reading for all those engaged in the building process. Drawing on data collected by research on real building projects it makes a number of interesting recommendations in relation to management and relationships between the parties. It also offers suggestions for the education and training of architects, including the view that the formal training of architects should include practical experience on site.

The way forward

Accordingly, it is not really productive to analyse whether particular contractual regimes are more or less likely to provoke disputes. The problems lie in a combination of the matters addressed in Chapter 2, exacerbated by conflict between those involved at the point where the opposing parties come into contact, or among members of the same team. The widespread criticism of management contracting as a breeding ground for litigation, caused by tyrannical construction managers and cussed contractors is, upon reflection, simply the product of large and complicated projects where insufficient care is taken over creating an atmosphere where the parties actually attempt to work together for their mutual benefit rather than allowing grievances to grow and fester to the ultimate detriment of the project.[3.1]

It is much less easy to provide a cure than it is to identify the problem. Some of the answers suggest themselves, the most obvious being the maxim that much of the success or failure of a project will be determined before the works on site begin. However it would be naive to suggest that planning and team selection were the be all and end all. Comments about the best laid plans are often apt. While it is impossible to suggest cures for the problems which occur, the remainder of this chapter attempts to offer some suggestions.

The Latham Report questions

Again, the Latham Report provides an interesting perspective. At paragraph 5.17 the author poses five questions:

(1) Are there too many forms of contract or too few? Does the number matter?
(2) Are some of them inherently adversarial, or likely to produce conflict because of the modern structure of the industry?
(3) Are there some procurement routes which are more likely to

produce a result which meets the client's wishes, and which should therefore be followed? If so, which?

(4) Are there some features which should be adopted across a wide range of contracts?

(5) Are there any contracts which should be used more often?

In answering his questions, Sir Michael Latham, in essence, offers a resounding 'no' to the first. In relation to the second he suggests that

- the separation of design and construction
- the pre-planning of all design work which is then not changed once the works have commenced
- the execution of the work by the contractor and not by domestic sub-contractors and
- the use of the same team both to administer and design the works

are all factors unlikely to meet the needs of the modern industry.

For the third, he recommends the use of mutually agreed modules to create flexibility and familiarity. As to the fourth, he suggests

- a general duty to trade fairly
- the creation of clearly defined work stages
- the agreement in advance of the price of any variations and
- the use of an independent adjudication system (a suggestion now implemented by section 108 of the Housing Grants, Construction and Regeneration Act 1996).

Finally, he suggests that the New Engineering Contract has much to commend it.

To a great extent the comments made in relation to his first three questions echo the points made above. They do not of themselves solve the problem of personality-driven conflicts, but they may mark a way of reducing them. The recommendations on the latter two represent a personal perspective. Widespread doubt has been expressed as to whether a duty to act fairly can or should be imposed. It is obviously desirable but, frankly, it is excessively optimistic to believe that it can be imposed where, after all, what is fair for one party may be unfair for another. The attractions of the

NEC are a matter of personal preference and much has been written about its novel approach. Certain drafting issues have been met in the second edition but time will tell how successful it proves in reducing the scope for strife. It is still too soon to know whether the fears are justified that have been voiced, namely, that its basic philosophy, and not just its drafting reveals major structural flaws rather than simply teething problems.'

Problem spotting

The difficulty with the problems which have been considered in the previous chapter and in the first part of this one, is that even if the parties have identified the existence of the problem, it will seldom be possible to assess accurately either its full extent or the probable consequences. This will be important when considering notice requirements. However, in many instances, the parties will argue with some cause that they simply did not see the events in question developing into the sort of delays which resulted. Without doubt those blessed with hindsight, generally the lawyers and claims consultants, will often suggest that the parties have closed their eyes to the inevitable and in so doing have done nothing to minimise or avoid the consequences.

It follows that if the parties translated their efforts in planning the works into analysis of the events and their effect upon the sequencing and resourcing of the works they will be able to react and reschedule the works so as to avoid delays. This is obviously a point for further consideration and is dealt with in Chapter 7.

However, in the real world, this will be described as a counsel of perfection. All too frequently, the reaction to the emergence of a problem is for the parties to take up polarised attitudes. Professor Eric Green has described the process of 'positional bargaining' where the parties to a dispute each take up an extreme stance designed to convey their position of maximum advantage.[3.2] From here, the combatants can either bargain their way down to a position where they meet somewhere in the middle or retreat to an impasse where both have become entrenched and the only way forward is litigation. The realisation that such problems need early treatment can be seen in the provision of the right to adjudication contained in section 108 of the Housing Grants, Construction and Regeneration Act 1996. The intent of this provision is to provide a mechanism for dispute to be dealt with quickly and efficiently in a way which enables the works to proceed before the parties become too entrenched.

The point is a simple one. Particularly when seen in the context of delay claims, small problems tend to develop into bigger ones unless caught early, hence the importance of the contractual system of notices as giving advance warning of anticipated difficulties and the need to monitor delays considered in detail in the rest of this chapter.

3.2 Notices

The requirement for the contractor to notify the architect of the occurrence or likely occurrence of delays is found in clause 25.2 of JCT 98 and clause 44(1) of the ICE 7th edition. In the NEC Second Edition, the matter is dealt with in clause 61.1. However, for present purposes it is worth simply setting out the critical words from each.

JCT 98 clause 25.2

1.1. If and whenever it becomes reasonably apparent that the progress of the Works is being or is likely to be delayed the Contractor shall forthwith give written notice to the Architect of the material circumstances including the cause or causes of delay and identify in such notice any event which in his opinion is a Relevant Event.

2 ... In respect of each and every Relevant Event
   ... the Contractor shall, if practicable in such notice, or otherwise in writing as soon as possible after such notice:

.1 give particulars of the expected effects thereof; and
.2 estimate the extent, if any, of the expected delay in the completion of the Works beyond the Completion Date...

ICE 7th Edition clause 44(1)

Should the Contractor consider that

(a) any variation ordered under Clause 51(1)or
(b) increased quantities referred to in Clause 51(4) or

(c) any cause of delay referred to in these Conditions or
(d) exceptional adverse weather conditions or
(e) any delay impediment prevention or default by the Employer or
(f) other special circumstances of any kind whatsoever which may occur

be such as to entitle him to an extension of time for the substantial completion of the Works or any Section thereof he shall within 28 days after the cause of any delay has arisen or as soon thereafter as is reasonable deliver to the Engineer full and detailed particulars in justification of the period of extension claimed in order that the claim may be investigated at the time.

NEC clause 61.1 and 61.3

61.1 For compensation events which arise from the Project Manager or the Supervisor giving an instruction or changing an earlier decision, the Project Manager notifies the Contractor of the compensation event at the time of the event. He also instructs the Contractor to submit quotations, unless the event arises from a fault of the Contractor or quotations have already been submitted. The Contractor puts the instruction or change decision into effect.

61.3 The Contractor notifies an event which has happened or which he expects to happen to the Project Manager as a compensation event if

- the Contractor believes that the event is a compensation event,
- it is less than two weeks since he became aware of the event and
- the Project Manager has not notified the event to the Contractor.

While the language used differs from that in either the JCT or ICE form, it is suggested that the effect is remarkably similar.

The other members of the various families of standard forms and the sub-contracts for use with them all include similar provisions adapted to suit the circumstances in which they are to be used. It will be clear at once that, although vastly different words are used, the principle in each is identical. Each places the onus on the contractor to provide the notice of delay either when the delay occurs when its effect becomes apparent or as soon as he can subsequently and the notice must identify the effect of the delay.

Practical questions

This will give rise to three practical questions:

(1) When will these notices actually be required in practice?
(2) What will the notice be required to say?
(3) What will happen if the notice is not served in accordance with the provisions of the various clauses?

Although it seems as though a respondent party will suggest almost as a matter of course that the claimant has not complied with the relevant notice provisions there is surprisingly little in the way of decided authority on these questions. In fact that authority is substantially confined to the justly famous decision of Mr Justice Vinelott in *London Borough of Merton* v. *Stanley Hugh Leach Ltd* (1985) 32 BLR 51. As is well known, the decision comprised an appeal against an interim award of an arbitrator in relation to a series of preliminary issues. These included a detailed review of the provisions of clauses 23 and 24 of JCT 63, the forebears of clauses 25 and 26 in JCT 98. For present purposes it is sufficient to say that, although the wording of the earlier clauses is very slightly different from that which has been quoted above, it is very difficult indeed to see how a significantly different meaning can have been intended by the draftsmen of the later contract. Hence, it is probably safe to conclude that, although *Merton* v. *Leach* is not binding authority in relation to cases on other forms of contract, it will be highly persuasive. It would take a brave tribunal to depart from it, in the absence of some clear indication in the words used to the effect that the parties intended that some other consequence would follow.

Mr Justice Vinelott began by dealing with Preliminary Issue 14 which asked whether the contractor was entitled to an extension of time in the event that he failed to give written notice of one of the causes of delay upon it becoming reasonably apparent that the progress of the works was delayed. In other words, is the notice a condition precedent to the right to an extension of time? Mr Justice Vinelott held that it was not. His reasoning can be summarised simply.

(1) He rejected the proposition that in the absence of a notice the architect had no responsibility to consider delays; and

(2) accordingly he found that, even in the absence of a notice, the architect still had a duty to consider delays and to award appropriate extensions of time; but

(3) acknowledging that the failure to give notice would constitute a breach by the contractor, he held that the contractor should not be entitled to any longer extension than he would have received had he served notice and thus not committed the breach.

In forming this opinion, which supported that of the arbitrator, he took support from the then current edition of *Keating*.[3.3]

Having decided this question, Mr Justice Vinelott turned to Preliminary Issue 6, which asked which of the various documents identified by Leach did in fact qualify as notices within the meaning of clause 23. As with the previously considered question, Merton argued that the clause should be construed strictly and that a document could not be a notice unless it specified a cause of delay in sufficient detail to enable the architect to form a view as to whether the cause of delay fell within one of the causes of delay specified in the contract, and if it did what delay would result or had resulted. The arbitrator rejected this view holding that the notice

'is simply to warn the architect of the current situation regarding progress. It is then up to the Architect to monitor the position in order to form his opinion.'

Mr Justice Vinelott concluded that the architect was entitled to expect the contractor to play his part, and that a failure to serve notice could be taken into account in assessing the extension of time granted. Adopting this approach he nevertheless held that, although the question of whether individual documents could in fact be notices or not was one to be referred back to the arbitrator, the question would be construed widely. In other words, the question was one of fact to be determined by a judge or arbitrator, but the presumption would usually be in favour of the notice being valid. Adopting the early warning criterion suggested by the arbitrator, arguments that a particular document cannot fulfil the requirements of the relevant clause are unlikely to succeed provided that the document does in fact refer to delay in some form. It then appears to be for the architect to pick up the trail.

Preliminary Issues 7, 8 and 10 overlapped. These dealt with whether the contractor owed a duty to particularise the loss and expense which would flow from a particular delay or whether this duty rested with the architect or at least that it was for the architect to instruct the quantity surveyor to ascertain that loss and expense. On behalf of Merton it was argued that since the contractor alone was in charge of planning the works it must follow that it was for him to identify the consequences which would follow from a failure to supply requested information or to serve proper notice. Mr Justice Vinelott allowed that this was an attractive argument but felt that it did not get the employer past the hurdle which had tripped him up previously, namely that the contractor's failure did not relieve the architect of his responsibility to ascertain delays and to instruct the quantity surveyor to ascertain loss and expense, but that if as a result of the contractor's failure to serve notice or details the architect awarded him less than he considered to be his due, he would really only have himself to blame.

Mr Justice Vinelott's judgment therefore proceeds from the essentially practical basis of regarding the notice provisions as providing protection for both parties, while not serving to relieve the architect from any of his own duties. Hence, the obvious lesson is that the important function of the notice is as a means to communicate, particularly in relation to events giving rise to delays, so what is important is not the form of the notice but what it says. In Appendix 2 there is a possible example of what might be covered in a notice of delay. This is put forward not as having any authoritative force but simply because it appears to fulfil the criteria set out by Mr Justice Vinelott and shows that such matters need not be complicated. Obviously in relation to a design and build contract the obligations which Mr Justice Vinelott found on the part of the architect will remain with the employer but in other respects the case will be equally applicable to design and build contracts. This is reflected in the second version in the appendix.

Reluctance to give notice

Even allowing for the comments by Mr Justice Vinelott, the notice requirements of most forms of contract give rise to a dilemma. Will serving a notice that the works are being delayed by an event which may give rise to a claim for an extension of time or to loss and expense serve to make what may already be a strained situation worse? The response to the question – 'Why didn't you serve a notice when it must have been clear that a delay was on the cards?' will often (quite understandably) be, 'We were anxious to avoid the situation where things "got contractual"'.

The difficulty with this answer is that it presupposes that there will be a collective wish on the part of all concerned to prevent this state of affairs from getting worse. If the employer sees that his project is running behind schedule and the contractor anticipates that without extensions or claims the project will lose money for him the hope that everything will come right in the end is likely to be a pipe-dream. Clearly, the purpose of adjudication is to enable such matters to be dealt with without delay during the works. Plainly, if a procedure exists whereby there is a quick and efficient means of dealing with disputes, it is less likely that such disputes will degenerate into litigation or arbitration and a probable break-down in the relationship.

The dilemma is therefore one which answers itself. If notice is served it will probably not make an awkward situation worse and if it is not served the problem and perhaps the dispute which results from it will still take its course. In fact there are almost invariably two arguments which militate against the 'get contractual' argument.

(1) Failure to serve notice will constitute a breach of the provisions of the contract. Even if this does not serve to debar the contractor from making a claim at a later date, it will provide the opposing party with ready ammunition with which to attack the claim. It will also inhibit the chances of successfully resolving the problems through adjudication.

(2) As will be apparent, it is the content of the notice rather than its form which is critical. There is no reason why the notice need be a confrontational document; it simply serves to articulate a problem. Experience suggests that, in a great many cases, the existence of a notice will give the parties a 'datum point' from which both can consider their respective positions. It may therefore serve to help resolve the dispute.

In some respects, the decision in *Leach* v. *Merton* is open to misinterpretation. It has been suggested in some quarters that it provides authority for the proposition that notices are unnecessary. This is not a view which will stand up to scrutiny. It is not what the judge actually held.

The benefits of giving notice

Certainly the perspective of those advising when disputes have arisen suggests that the cases where the parties have failed to serve appropriate notices are far harder to settle than those where the problems are clearly mapped out by way of notices. As will be apparent from the next section, the provision of proper notices is an invaluable aid to plotting the delays which have occurred. However, it is more than just a case of developing good habits. The position has been summed up by John Riches of Henry Cooper Consultants:[3.4]

'The problem is that if I am asked to prepare a claim and my client has failed to serve the proper notices, it becomes that much harder to produce a claim which will actually convince someone that the things you are complaining about actually did affect the project in the way you say they did. How can you say that something was really a critical delay to the whole project when at the time nobody has even seen fit to write a letter about it.'

Turning this around it is not difficult to see the underlying purpose behind the requirements in the principal standard forms. Even in the New Engineering Contract, where it is clear that in coining the terms 'delay events' and 'compensation events' the intention has been to escape from the conflicts which were seen as inherent in the wording of the JCT forms, the point in having a notice clause is to act as a an early warning system to alert the parties to some occurrence that had not been anticipated at the time when the works were planned, and to act as a trigger to enable the parties to assess its effects and make provision for them. Accordingly, failure to comply with these requirements (deliberate or otherwise) deprives the parties of this opportunity and is likely to have the precisely opposite consequences to those which might have been intended.

Again, while most of the above has looked at the matter from the contractor's perspective, it is equally applicable to that of the employer. This is illustrated by the case which was considered in another context earlier in this chapter, *West Faulkner v. London Borough of Newham*. Here, despite the manifestly poor performance of the contractor, the architect had failed to issue a notice pursuant to the terms of clause 25(1) of JCT 63 (now 27.2 of JCT 98) to the effect that the contractor was not proceeding 'regularly and diligently' with the works. It was contended on behalf of the architect that in order to meet the requirements of this clause it had to be shown that the contractor was proceeding neither regularly nor diligently and that, although there could be no doubt that he was not proceeding regularly, nonetheless he was doing his best to proceed diligently – although he was not making a very good job of it. The contention was rejected emphatically by the Court of Appeal. More significantly the architect's failure to serve notice was the effective cause of a substantial claim by the employer against the contractor and subsequent proceedings against the architect resulting in a finding of professional negligence against him. The message speaks for itself.

3.3 Monitoring delays

This is the next rung up the logical ladder from serving or receiving notice of delay. The notice itself is often no more than an early warning signal (to use Mr Justice Vinelott's expression in *Leach*) and is sometimes understandably a fairly crude instrument. If the contractor seriously intends to advance a claim for an extension of time he must anticipate that the employer will ask him what the precise effect of the notified event is likely to be. The employer would be ill-advised to ask this question unless he has some idea of the likely answer. Hence the issue on both sides becomes one of determining precisely when, why and how long?

It is stating the obvious to say that this task is easier if the parties have the information to determine not only what happened but also why, when it happened relative to other events, how long it took and its impact upon other activities. It is obviously easier to assemble that information during the currency of contract than to attempt to recreate the story after the events have long since occurred. While the comments in the preceding section on notices are part of that story, they are only a facet of that task.

Record keeping and correspondence

Again, it is a truism to say that there is no substitute for good record keeping, the maintenance of comprehensive site diaries, confirming instructions, notifying the release of or failure to supply drawings and design details, and keeping correspondence up to date, together with proper adherence to the contractual requirements (whatever they may be) regulating and updating of programmes for the execution or planning of the work. The first part of this chapter dealt with the ways in which conflict may arise on site. It is probably worth repeating the view that disputes are more frequently caused not by sinister attempts to exploit differing bargaining strengths but by poor organisation and a failure to appreciate the obligations which the contract imposes upon the parties.

It is easy for lawyers and claims consultants to wag admonishing fingers in the face of failure to keep proper records, but the unfortunate fact is that these things are frequently symptoms of a deeper problem with the planning of the works – a neglect either at the tendering or mobilisation stage to ascertain what the particular project is likely to involve. In recessionary times this will prompt the response that the resourcing of works at the same levels which prevailed in times of comparative prosperity is not possible for a contractor who actually wants to be successful with any tenders. There is no complete answer to this. However, the comment might be made that, even in times where many parties' resources have been pared down to the minimum, a consideration of the facts simply does not support the 'well, I only had one pair of hands' sort of argument.

This is sadly reflected in a great many of the projects which end in the clutches of lawyers and is shown by the wealth of occasions where the parties have 'descended into the arena' in protracted and tendentious correspondence. Inevitably it is easier to say than to do but it may be sensible to ask whether, for example, it is not more sensible to use limited resources to write three short clear letters notifying the architect of a failure to supply specific pieces of information rather than one long vague one, followed up by arguing the toss in two further letters replying at length to his contention that the first is either not a proper notice or that the information was late. Put another way, the effect of writing argumentative letters is usually only the production of confrontational responses.

Nowhere is this illustrated better than in the minutes of many typical site meetings which serve more to reflect the extent of the degeneration of interpersonal relations on site than to fulfil the more useful function of acting as a record of the delays which a project has actually suffered.

The purpose of the records

The purpose for which records are kept is initially to determine what exactly occurred. The related issues of how these events interlinked will only be possible once it is ascertained what actually happened. To do this it is necessary to have some starting point against which to monitor delays. In the simplest contracts this will comprise no more than a start and end date. An example is shown in Fig. 3.1. This comprises a simple task composed of one activity. In scenario No. 1 the event has occurred during the works causing the end date to be pushed back and in scenario No. 2 it has happened before the start, again pushing back the end date.

In the majority of construction projects the question of monitoring delays will involve a good deal more than the consideration of a single event. However it is a starting point, particularly when we remember that at this stage we are not concerned with why these things occurred. Accordingly, we can expand the single event model into a slightly more plausible model (Fig. 3.2). Here, the project comprises six separate activities. Again, the events shown in the footnotes have simply been plotted on to the barlines for the activities – we have not considered why they have happened or what their relationships or restraints might be.

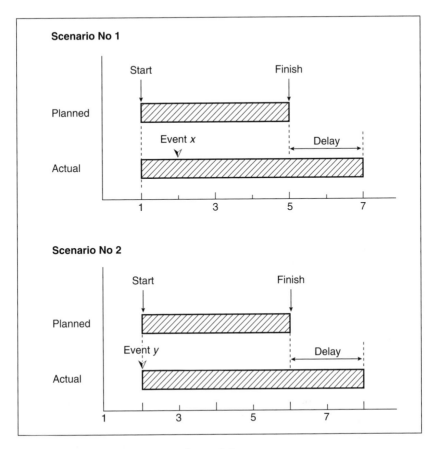

Fig. 3.1 Delays: a more simple model.

Reprogramming

This then leads to two separate issues for consideration. What should the delays be monitored against and how do we then go about analysing those delays by reference to the restraints and dependencies which they bear upon one another. The latter will be considered in Chapter 8. The former calls for further consideration of the purpose of programmes. What happens when the 'contract programme' considered in the previous chapter ceases to be relevant to the remaining works. Regrettably, there is no simple answer. None of the commonly used standard forms deal comprehensively

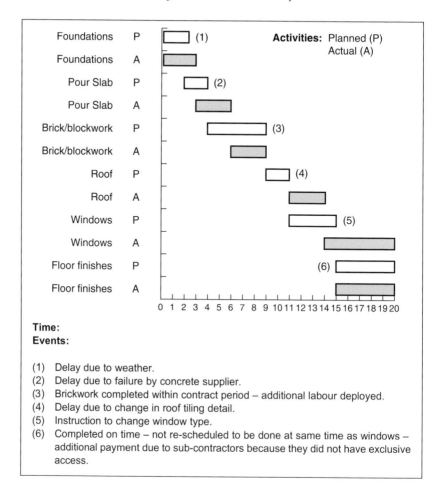

Fig. 3.2 Delays: a more complex model.

with the question of reprogramming. To fill this gap, bills of quantities and Employer's Requirements have increasingly sought not only to impose obligations in relation to the production of the contract programme but also to require that it should be updated and modified to suit the changing nature of the works.

In short it is impossible to put forward a generally applicable rule. However, while there will be as many different scenarios as there are different contractual obligations, it is possible to set out a few broad categories.

Reprogramming

The first is that covered by the sort of preliminary provision set out in Chapter 2 and reproduced in Appendix 1. The contractor is obliged to produce a programme (which he is also entitled to work to) and to update it to suit changing circumstances. It follows that the contractor is entitled to measure delays against whichever programme is then current and reflects the contractor's present obligations in relation to the remaining works. Expanding the model set out in Fig. 3.2, this will produce the situation where event 1 causes the contractor to produce a further programme to reflect the change in circumstances. Event 2 then occurs and the contractor can monitor the resultant delay against the completion date and sequence of working set out in the second programme. This is shown in Fig. 3.3.

Under the standard forms

The second is the situation encountered in the typical unamended standard form JCT contract where the contractor has produced a programme which he is probably entitled (but may not be obliged) to work to. There is no obligation to reprogramme. Hence, taking our example, even if the contractor has taken the sensible decision to produce a second programme, he is still entitled to plot the impact of the delays on the original programme. It is here that the distinction between what happened and why it happened and its consequences will demonstrate itself. The period of delay will be the same in each instance. However, the preceding and succeeding events will influence and be influenced by this matter and this will ultimately dictate the extent to which this event may delay the completion of the works. The task of unravelling the actual events and determining their interaction is, however, a matter for a later chapter.

Beck and call

The third is what we can call the 'beck and call' situation, following the *Kitsons* v. *Matthew Hall* (1989) 47 BLR 82 case which was considered in the previous chapter, at the end of Section 2.5. It will be remembered that Judge Newey had held that although the parties had produced a succession of programmes during the works, these were no more than a pragmatic way of fulfilling the

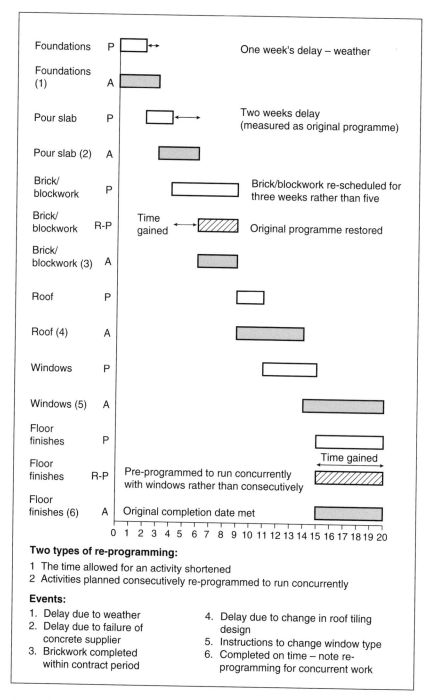

Fig. 3.3 The programming.

contract and that the requirement that the plaintiff should work in accordance with the dictates or requirements of the defendant meant that the defendant could issue instructions as he thought appropriate which could alter the order in which the plaintiff carried out the works, and the plaintiff's only obligation was to complete the works by the completion date.

In this situation, which is considerably more common than might be thought, the programme is of little help in monitoring delay. The only yardstick is the completion date. However, unless the contractor can show that the effect of an instruction was to render the works incapable of performance within the contract period, the prospect of successfully claiming an entitlement to an extension of time is bleak. In practical terms, this will generally rule out anything which alters the sequencing of the original contract works since the contractor will still face the argument that his only entitlement was to be able to complete within a specific period and that he had no right to require that any particular works would be made available at any specific date, whether provided in the programme or not. Time based claims are thus likely to be confined to variations in the works themselves – physical variations.

Time at large

The final scenario is the 'time at large' argument. This has already been considered in Section 2.5 in relation to the line of old cases starting with *Bottoms* v. *York Corporation* (1892) 2 HBC 208 and *Thorn* v. *City of London* (1876) 1 App Cas 120. While this argument has proved very popular with the authors of claims, it will only really be available in the most extreme instances. Unless it can be shown that the whole of the contractual mechanism governing extensions of time has broken down, it will still be for the claiming party to say that he is entitled to measure delay against the contract programme, or perhaps a revised programme which governs the sequence in which he was entitled and obliged to work.

3.4   Conclusions

The lesson to be learned from this is simple. The method to be used in calculating delays will vary according to the contractual regime by which the works are to be executed. This of course means that the likelihood of one party sustaining a delay claim will be another of the features to be addressed by the parties in negotiating the original contract and thus allocating risk. What is equally clear is that except in the rarest of instances it will be difficult if not impossible for the parties to throw away that contractual machinery when attempting to analyse both the delays and responsibility for them. The importance of this will become clear from later chapters but it is sufficient for the present to note that the monitoring of delays is something which can only really be done effectively by reference to the actual events which occurred as the works progressed.

We can therefore draw together the three strands upon which this chapter has concentrated:

- the difficulties inherent in contractual procurement, and the apportionment of risk which this involves;
- the need to comply with the contract requirements regulating performance and dealing with the giving of notice; and
- the difficulties associated with monitoring delays when they occur.

It follows from this that 'getting it right', which for these purposes means avoiding conflict, will be achieved by successful negotiation of the three.

CHAPTER FOUR
COMPLETION DATES

4.1 Introduction

Standard form construction and civil engineering contracts generally use a two stage approach to determining when the works are complete, the certificate of practical completion and the certificate of making good defects. This involves an initial certificate or statement that the works are complete, followed by a 'maintenance period', during which faults emerging in the newly completed works are addressed. This culminates in a further certificate that defects have been put right, and a final certificate when, in theory at least, all outstanding issues are dealt with. At the time of the initial certificate it is customary for half of the retention which has been deducted from interim payments during the works to be released. The remaining portion or 'moiety' is released at the end of the defects period. This is intended to act as the incentive to the contractor to deal with defects and thus as protection for the employer.

Identifying the date of completion

Under the JCT family of contracts, the initial certificate is referred to as the certificate or statement of Practical Completion and under the ICE family as Substantial Completion. Under the NEC it is referred to simply as Completion. The relevant clauses are clause 17.1 of JCT 98, clause 48(2) of ICE 7th Edition and the definition in 11.1(13) of the NEC. It will be seen that while the wording differs, the effect of the clauses is much the same under the JCT and ICE forms – the Supervising Officer or, in the case of design and build contracts, the employer or his agent issues a certificate or statement that the works are (Practically or Substantially) complete and this will be the date by which all issues under the contract relevant to the question of

(Practical or Substantial) completion of the works will be computed.[4.1] Under the NEC the position is much the same although the language is simpler – the contractor does the work so that Completion is reached on or before the Completion Date and the project manager issues a certificate to that effect (clauses 30(1) and 30(2)).

Effect of the date of completion

This date is therefore critical. The date of Practical Completion will be the date by which delays are measured. Extensions of time will be sought, calculated and granted by reference to this date. Correspondingly, the entitlement to deduct liquidated and ascertained damages will be triggered by the failure to achieve Practical-Substantial Completion by the date stated in the contract.

4.2   Defining completion

The expressions 'Practical' and 'Substantial' completion[4.2] are not defined precisely in either the JCT or ICE forms. Both expressions might be thought to connote the situation where the works are 'complete for practical purposes', or 'substantially, as in "nearly" complete', or 'complete but for minor or insignificant matters' and so on. Indeed where the building works form part of a larger transaction in which the employer under the building contract has also entered an agreement with a forward purchaser or a funder it is common to see Practical Completion defined as 'complete but for *de minimis* or snagging items'.

There is surprisingly little authority on this point. The leading authority is that of *J Jarvis and Sons Ltd* v. *Westminster Corporation* [1970] 1 WLR 637. In the Court of Appeal Lord Justice Salmon had favoured the 'complete for all practical purposes' approach but in the House of Lords Viscount Dilhorne said

'One would normally say that a task was practically completed when it was almost but not entirely finished, but "Practical Completion" suggests that that is not the intended meaning and what is meant is the completion of all the construction work that has to be done.'

The matter was, however, considered by Judge Newey in *HW Nevill (Sunblest) Ltd* v. *William Press & Sons Ltd (1982)* 20 BLR 78. Judge Newey's view favoured the 'complete but for *de minimis* items' interpretation of the expression 'Practical Completion'. He stressed that if there were any patent defects in the works, then the architect could not properly certify Practical Completion. Unfortunately, this view is not developed – the judge does not define either *de minimis* items or patent defects. It does, however, accord with the ordinary meaning of the words. The issue came before the courts again in *Emson Eastern Ltd* v. *EME Developments Ltd* (1991) 55 BLR 114, another decision of Judge Newey. Although Judge Newey observed that the view he had expressed in *Nevill* was '*probably* right' he had previously suggested that there was no room in the JCT contract for defining 'completion' differently from 'Practical Completion'. For the purposes of triggering an entitlement to deduct liquidated damages this is almost certainly sensible, and this, rather than the definition of Practical Completion, was primarily what this case was about. It does, however, leave the lingering doubt as to what discretion an architect has to overlook minor or snagging items when deciding that the works are practically complete.

4.3   Calculating the delay

In the simplest contracts the contractor will carry out the works and the employer will pay the agreed price. The effect on such contracts of delay is dealt with in Chapter 2. In modern standard form contracts, the parties agree not only a completion date but also a mechanism for extending the contract period in the event that the works are delayed. In all of the standard forms the operation of this machinery is triggered by the occurrence (or non-occurrence) of Practical Completion. If Practical Completion is not reached by the date stipulated in the contract this is necessarily because of occurrences causing delay – whether on the part of the contractor or the employer. No doubt if events have occurred during the course of the works which have caused losses to the parties, these may have a legal consequence, but they will not found a delay claim unless the contract period has been exceeded.[4.3]

This leads to the question of calculation of the delay. Applying the wording of the standard form contracts, the delay will be the period between the date for completion provided in the contract as extended by the machinery of the contract and the date when Practical Completion of the works was actually certified. Two conclusions follow from this. Firstly that delays will be calculated by reference to the date fixed in the contract for completion of the works and secondly that until the date for completion has been passed there will be no delay.

In a number of instances, however, attempts have been made to calculate the period of delay differently. Firstly, there are claims where it is suggested that while the actual period of delay amounted to x weeks, the delaying events were such that the contractor would have been delayed by $x + y$ weeks but for the efforts he took to speed up his performance. This approach is considered further in the contexts of acceleration (Chapter 5) and calculation of entitlements to extensions of time (Chapter 8). It has tended not to be an approach which has found favour with the courts because of the difficulties which it presents in proving what actually happened and the inter-relationship between events.

The difficulties are demonstrated by looking at a number of cases where such attempts have been made. In each of these cases it will be seen that ingenious arguments were formulated which the courts rejected. In *Glenlion Construction Ltd* v. *The Guinness Trust* (1987) 30 BLR 89 the court determined that the contract programme entitled the contractor to complete the works prior to the date fixed in the contract for completion. However, that did not give rise to a corresponding obligation on the part of the employer to do everything necessary to allow him to do this. In other words, the contractor could not argue that an act of prevention by the employer which absorbed part of the period between the programmed date and the later contractual completion date would allow him to claim extensions of time for that period. In *Peak Construction* v. *McKinney* (1970) 1 BLR 114 the Court of Appeal had held that it was impossible to hold that a delay of 58 weeks could be attributed to remedial works which had taken only six weeks to complete.

The point was considered in a slightly different context in *Balfour Beatty* v. *Chestermount Properties* (1993) 62 BLR 1. It was argued on behalf of the contractor that, in circumstances where an instruction had been issued after the date for completion, the effect of that act was that the extension ought to be calculated from the date of that instruction. The employer contended that the right approach was to look at the effect of the instruction and add the effect of that to the existing contractual completion date. The difference between the two is illustrated on Fig. 4.1. The contractor's approach was referred to by Mr Justice Colman as the 'gross' method of calculation and the employer's as the 'net' approach. The judge looked at what he called the 'underlying realities' of the situation. He concluded that the proper meaning of the contract was that it required extensions to the contract to be measured by reference to what had actually happened. Considerable weight was placed on the approach taken in *Peak* v. *McKinney*. Hence, if an event had occurred which caused a particular period of delay, that was the extension which should be granted. It was unrealistic and arbitrary therefore to take the gross approach which was likely to result in extensions which had no relationship to the actual period of delay that had been caused by a particular act.

Accordingly, the period of delay will be the difference between the date for completion stated in the contract and the date when the works are actually completed.

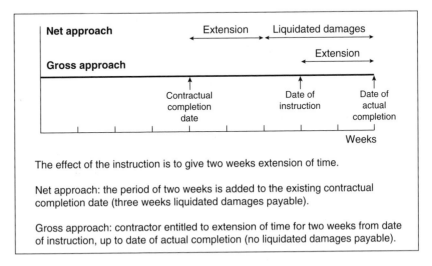

The effect of the instruction is to give two weeks extension of time.

Net approach: the period of two weeks is added to the existing contractual completion date (three weeks liquidated damages payable).

Gross approach: contractor entitled to extension of time for two weeks from date of instruction, up to date of actual completion (no liquidated damages payable).

Fig. 4.1 Net and gross approach.

4.4 Certificates of non-completion

The duty of the contractor is to complete the works by the date
stated in the contract. Put at its simplest, the failure to complete the
works by the completion date will give rise to an entitlement on the
part of the employer to deduct liquidated and ascertained damages
at the rate appearing in the contract. The form of the liquidated
damages provisions appearing in the JCT family of contracts differs
significantly from that in either the ICE or GC/Works forms. Different conditions precedent apply to the right to deduct liquidated
damages and, in default of compliance with those strict conditions,
the deduction of liquidated and ascertained damages is unlawful.
The decisions of the courts dealing with questions of compliance
with those conditions precedent (e.g. *JF Finnegan* v. *Community
Housing Association* (1995) 47 CLR 25) have made it clear that if a
particular contract form insists upon certain preconditions being
met in order to allow the deduction of liquidated damages, those
conditions will be enforced strictly.

The different provisions are set out and discussed comprehensively by Brian Eggleston in *Liquidated Damages and Extensions of
Time*. For our purposes, the important rule is that until the completion date or extended completion date in the contract has been
passed there can be no delay, and neither can there be any entitlement to liquidated damages. It also follows that without an event
causing or likely to cause delay there can be no entitlement to an
extension of time. In all of the standard forms it is provided that the
granting of an extension of time will automatically postpone the
employer's right to claim liquidated and ascertained damages for
the period of the extension. The contractor's entitlement to additional payment, by contrast, will depend upon how the parties to a
contract have chosen to apportion particular risks.[4.4]

The inter-relationship between liquidated and ascertained
damages and extensions of time has important tactical ramifications. Frequently, the principal reason for claiming an entitlement to
more time is directed towards defeating a claim for liquidated and
ascertained damages. This is the most frequently encountered form
of 'defensive' claim, an idea considered in more detail in Chapter 5.
It also explains the proliferation of claims for extensions of time
based on a variety of different relevant events, some entitling the
contractor to additional time and money, others merely further time

but still relief from a claim for liquidated damages. While it is easy to see the sense in adopting a fall back position, it will be seen in Chapters 5 and 6 that this may well provide difficulties to the claimant who will have to prove alternative causes for the delays which he has suffered – and runs the risk that by attempting to construct two arguments he lessens the impact of both.

4.5   The duty to review

Keating rightly observes that it is desirable that extensions of time should be granted to a date in the future so that the contractor can plan his work accordingly. The bonus which this provides is that it may enable the parties to programme the remainder of the works in such a way as to avoid further delays. It follows from this that it is incumbent on the contractor to give notice of delays timeously.

Clauses 25.2.1.1 and 25.2.2 of JCT 98 have been considered in the previous chapter. Almost identical provisions appear in WCD 98.

Clause 44(1) of the ICE 7th Edition is similar providing that the contractor

'shall *within 28 days* [emphasis added] after the delay has arisen or as soon thereafter as is reasonable in all the circumstances, deliver ... full and detailed particulars in justification of the period of extension claimed in order that such claim may be investigated at the time'.

Substantially the same wording appeared in the 6th and 5th Editions.

However, while the standard forms recognise the need to give prompt notice of the occurrence of delays, it is recognised that it may not be possible fully to assess the effects of the delaying events at the time when they occur. Hence the review provisions which appear in each of the standard forms.

The JCT provisions

JCT 98 (both traditional and design and build forms) provides by clause 25.3.3. that:

'After the Completion Date, if this occurs before the date of Practical Completion, the Architect [or the Employer in the case of WCD 98] may, and not later than the expiry of 12 weeks after the date of Practical Completion, shall in writing to the Contractor . . .

> .1   fix a Completion Date later than that previously fixed if in his opinion the fixing of such later Completion Date is fair and reasonable having regard to any of the Relevant Events, whether upon reviewing a previous decision or otherwise and whether or not the Relevant Event has been specifically notified by the Contractor under clause 25.2.1'

and sub-clause 25.3.3.2 provides a corresponding power to fix an earlier completion date.

> 'having regard to any instructions issued after the last occasion on which the architect fixed a new Completion Date.'

The power to review is clearly discretionary although the architect is obliged to conduct a review and fix a new date or confirm the one previously fixed. The contractor cannot complain if the architect decides that there is nothing to review. However, in exercising his discretion the architect is obliged not only to have regard to matters which have been notified to him, but also to matters which have not. Under sub-clause 25.3.3.1 the architect can review all of his previous decisions to ascertain whether any further extension should be granted. By contrast, the power to fix an earlier completion date under sub-clause 25.3.3.2 is limited to matters which occurred after his last review. The conclusion is that once granted an extension of time cannot be undone.

The ICE provisions

Clause 44 (5) of ICE 7th Edition provides that within 28 days of the issue of the certificate of Substantial Completion of the works the engineer shall

> 'review all the circumstances of the kind referred to in sub-clause (1) [circumstances giving rise to delay] of this Clause and shall

finally determine and certify to the Contractor the overall extension of time (if any) to which he considers the Contractor entitled ... No such final review of the circumstances shall result in a decrease in any extension of time already granted by the Engineer.'

Clause 44(5) of the 6th Edition was expressed in identical terms, but required that the review should be carried out within 14 days of the issue of the certificate. Like JCT 98, the duty of the Engineer extends to matters not notified to him as well as to those which have.

Practicalities

Clause 25.3.3. was considered in *Balfour Beatty* v. *Chestermount* (a case which has been considered above in a different context). It will be recalled that the contractor had contended that the effect of an instruction issued after the date previously fixed for completion was that the completion date had to be extended *from that date* as opposed to being calculated by reference to the net extension of time to which the contractor was due, having regard to all previously granted extensions. In the context of the power to review it was contended that on a proper construction of clause 25.3.3, the power to review could only be exercised to grant a new completion date at a *future* date. Mr Justice Colman rejected this proposition, and held that the duty was to review the net extension to which the contractor was due, and that this could in many instances result in the completion date being fixed at a date *prior* to the date on which the review had taken place.

It is submitted that there is no reason why this should not be equally applicable to the provisions of clause 44(4) of ICE 5th Edition or clause 44(5) of ICE 6th and 7th Editions. Interestingly, clause 63.3 of the NEC Second Edition provides explicitly that delay is to be measured using the net basis.

A final point, and one not addressed in *Chestermount*, is that of the contractual origin of the event in respect of which the extension is granted. It is easy to envisage circumstances in which the architect or engineer might, upon reviewing the extensions previously granted, decide that, while they had been originally granted in respect of an event entitling the contractor to reimbursement, upon review the cause of the extension should be altered to one which did not. None of the standard forms address this situation. There is no reason in principle why the architect or engineer should not do this. However the consequences of a decision to alter the cause of an extension of time may well be just as serious as whether or not the extension is granted.

It is also likely that this is precisely the sort of situation which might usefully be addressed by adjudication. A quick and informed mechanism designed to produce a solution to disputes during the currency of the works is likely to have widespread application in these circumstances.

4.6 *The Final Certificate*

The review process discussed in the previous section is not intended to be a conclusive determination of the parties' rights to extensions of time. JCT 98 provides for the issue of a Final Certificate under clause 30.7 and under clauses 30.8 and 30.9 describes what the word 'final' is intended to mean. This includes the provision that the Final Certificate is intended to be conclusive as to any extensions of time granted, unless proceedings are commenced within 28 days. In fact this part of clause 30 has given rise to a good deal of debate and much litigation. As a result, this is a part of the standard form which is frequently amended and in practice it is commonplace for a Final Certificate not to be issued.

This is because the effect of a Final Certificate will be to close the door on the possibility of further extensions of time being granted. Conversely, it may have the effect of terminating the contractor's liability for liquidated damages. Accordingly, in projects where the parties appear likely to embark upon disputes as to the extensions of time which should have been granted, it is rare to encounter a Final Certificate.

This problem does not occur under the 7th Edition of the ICE form. Clause 60 (4) deals with the contractor's final account and the steps to be undertaken prior to the final payment. The expression 'final certificate' is not used. Neither is it stated in the clause that the final payment will be held to be conclusive of anything in particular.

CHAPTER FIVE
CLAIM PREPARATION: PRELIMINARY CONSIDERATIONS

5.1 Objectives

The purpose of this chapter is to look at a number of points which need to be addressed at the time when one of the parties takes the decision to assert an entitlement to an extension of time. These are points which will apply to some projects to a greater extent than others. They are of necessity a disparate list of factors.

Claims do not exist in isolation. As important as the question 'What are we entitled to?' is the question 'What are we trying to achieve?'

For contractors and sub-contractors the answer to these questions is usually that they are seeking to show

- that the works were delayed for a specific period as a result of specific causes and
- that in consequence of this they have incurred particular losses
- which they are entitled to recover.

Frequently this will be wrapped up with the contractor's need to establish an entitlement to an extension of time, which will defeat the employer's claim to be entitled to deduct liquidated and ascertained damages; or the sub-contractor's need to show that liquidated damages deducted from the contractor cannot be passed down the contractual chain. This chapter attempts to put that into a practical context and in the next we will look at the legal considerations which need to be addressed in the preparation and presentation of claims.

For the employer, the exercise is more often a defensive one in which the first objective is to justify the entitlement to retain liquidated and ascertained damages or to defeat claims for further time and loss and or expense. In a minority of cases, however, (*Wharf Properties* and *ICI* v. *Bovis* being good examples[5.1]) employers have sought to recover the additional costs and programming overruns suffered on a particular project, claiming that the delays have occurred consequent upon the failures of the contractor and the professional team, or, as in *Darlington BC* v. *Wiltshier Northern Ltd* (1994) 41 CLR 122 that the sums paid in excess of the original contract sum fall to be reimbursed as monies paid under a mistake of fact.

In many instances this will be linked to the need to open up, revise or review certificates of the architect, engineer or supervising officer. Until very recently, this also dictated the choice of forum. Prior to the decision of the House of Lords in *Beaufort House Developments (NI) Ltd* v. *Gilbert-Ash (NC) Ltd* (1998) 14 Con LJ 280 the need to open up a certificate of the supervising officer meant that the case necessarily had to be dealt with by way of arbitration because, in the wake of *Northern Regional Health Authority* v. *Derek Crouch Construction Co Ltd* (1984) 26 BLR 1, the courts were said to lack jurisdiction to deal with such matters. *Gilbert-Ash* changes that.

A typical scenario is set out in Fig. 5.1. It will be seen that the 'events' are common to each party and the reasons are the mirror image of each other and that the 'objectives' are mutually exclusive. Thus if the contractor succeeds in showing that his claim for an extension of time is justified, the employer's claim to liquidated and ascertained damages will almost necessarily fail.

This brings us back to the question of objectives. The question 'What are we really trying to achieve?' may in reality have several answers, including:

objectives	events	contractor's assertions	employer's assertions
extension of time	delayed start	late access	failure to mobilise
extension of time	delayed information	late drawings	contractor not ready
liquidated damages	late completion	not entitled to deduct	contractor culpably late

Fig. 5.1 Setting out objectives

- Using the claim as a device to bolster a party's negotiating position in relation to measured work or variations.[5.2]
- Defeating a claim for liquidated damages from a party higher up the contractual chain.
- Passing claims from the sub-contractors to the main contractor or employer, and perhaps in so doing obscuring the fact that those claims result from one's own poor performance.
- Delaying or complicating the settlement of claims by other parties in the contractual chain.

These are in addition to the obvious goal of achieving an extension of time. In these circumstances, the claim is frequently no more than a makeweight, advanced without any serious belief as to its validity.

The historian Clausewitz famously described war as diplomacy carried out by other means and, quite obviously, the use of claims for purposes other than the advancement of a clear legal right is akin to this. Such attitudes are often decried as indicative of a confrontational approach. Inherent within Chapter 8 of Latham is the endorsement of procurement methods based upon 'partnering' and the condemnation of business methods seen as leading to conflict and, by implication, claims.[5.3]

However, it is a mistake to regard Latham as an outright attack on the use of claims at any time. At Paragraph 9.4 he states that 'disputes may arise despite everyone's best efforts to avoid them'. This is important. The sort of approaches described above are often employed in an unscrupulous way. The object of the Latham Report and the Housing Grants, Construction and Regeneration Act 1996 is to reduce and outlaw the use of devices aimed at obstructing legitimate rights to payment with contractual claims that have no real foundation (see Chapter 10). The provisions of the Act requiring payments to be made by specific ascertainable dates and the availability of adjudication to enable disputes to be dealt with quickly and expediently is unashamedly intended to outlaw the use of such claims.

5.2   Example in practice

It is a mistake to believe that the post-Latham and Egan construction industries will be free from claims. What may, however, be helpful is to look at the example of a fairly simple construction project and analyse the points which may go wrong and the opportunities which the parties may have to put this right, taking some of the lessons from the previous chapters.

The purpose of this is not only to look at how these matters may be avoided but to look at the considerations which will need to be taken into account in preparing claims should it prove impossible to achieve settlement.

A very simple starting point is set out in Fig. 5.2. The problem arises out of a contract for refurbishment of a house carried out by a single contractor under the direction of a contract administrator. The diagram spells out the overly familiar series of problems. Taking this information, it is not too difficult to take an educated view as to the difficulties which the parties to these claims may face. That view can be refined by closer analysis of what the contractor and the employer will each have to demonstrate. This is set out below in Fig. 5.3. The acid test is what the parties can actually prove.

It will be noted that the expression 'cause and effect' does not appear in any of this example. Neither is there any sense in which this example relies upon sophisticated legal reasoning. Instead, this example depends entirely upon consideration of what the respective parties can actually prove. This is shown by comparison between the significant problems – the employer's various instructions for additional works and the changes in the electrical layout. The former plainly grabs the attention. Numerous instructions were given. Clearly they caused difficulties to the contractor and had the effect of knocking him off his stride. Volume, however counts for little in circumstances where the contractor can do no better than show that some extra payments are due in respect of the value of variations. No records, diaries, or correspondence can be produced which actually show what effect these matters had on the progress or timing of the works. At best, it may be that at a future date the contractor will be able to adduce the evidence of an expert to show that the matters under consideration would cause particular delays and to offer a view as to what those might be. At best, this is a speculative and theoretical exercise which may or may not bear a relationship to the events which occurred. Consideration of the *McAlpine Humberoak* v. *MacDermott* (1992) 58 BLR 1 and the *Bernhards Rugby Landscapes Ltd* v. *Stockley Park Consortium Ltd* (1997)

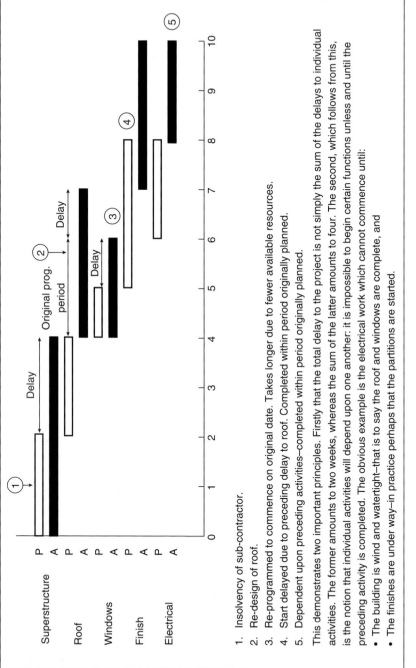

1. Insolvency of sub-contractor.
2. Re-design of roof.
3. Re-programmed to commence on original date. Takes longer due to fewer available resources.
4. Start delayed due to preceding delay to roof. Completed within period originally planned.
5. Dependent upon preceding activities—completed within period originally planned.

This demonstrates two important principles. Firstly that the total delay to the project is not simply the sum of the delays to individual activities. The former amounts to two weeks, whereas the sum of the latter amounts to four. The second, which follows from this, is the notion that individual activities will depend upon one another: it is impossible to begin certain functions unless and until the preceding activity is completed. The obvious example is the electrical work which cannot commence until:

• The building is wind and watertight—that is to say the roof and windows are complete, and
• The finishes are under way—in practice perhaps that the partitions are started.

Fig. 5.2

82 BLR 39 decisions (the latter considered below) show that this approach is frequently subject to strong judicial criticism.

By contrast, the electrical conduits issue is clear. A note exists ordering the change. Clearly, they were delivered five weeks late. Furthermore it is accepted that the installation of these conduits was the last task to be undertaken, apart from preparation to offer the works to the employer.

The point is simple. The fact that events have occurred which have hampered the parties do not of themselves give rise to any entitlement. This can also be seen from the decision of Judge Humphrey Lloyd in *Bernhard's Rugby Landscapes*. This case has already been considered in Chapter 3 but it is also teaches important lessons in relation to the objectives which must underpin any claim. The defendant's complaint was that a particular section of the defendant's claim (which the judge had already disparagingly referred to as a 'forest pleading' in which the statement of claim merely provided a prologue to the incorporation of a large and in parts impenetrable claim document) was oppressive to the defendant because

> 'it was not clear which variations or other causes of delay were relied upon as causing delay, how those variations correlated, ... what events (whether variations or not justified an entitlement to an extension of time under the contract and how much time was claimed for each event.'

Upholding this objection the judge directed that the plaintiff should state clearly how the variations were supposed to interlink and to what purpose.

A further point is that it is necessarily simpler to show the consequences of individual identifiable matters – such as the electrical items in Fig. 5.3 – than to allege that the cumulative effect of a group of matters had any particular consequence: see the items in Fig. 5.3 for additional work.

5.3 Means and ends

The basic requirements

In Chapter 3 we have looked at record keeping and in Chapter 4 at objectives. At the beginning of this chapter the questions 'What are

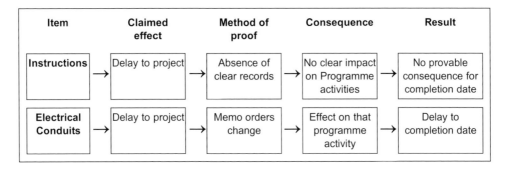

Item	Claimed effect	Method of proof	Consequence	Result
Instructions →	Delay to project →	Absence of clear records →	No clear impact on Programme activities →	No provable consequence for completion date
Electrical Conduits →	Delay to project →	Memo orders change →	Effect on that programme activity →	Delay to completion date

Fig. 5.3   Consequences of delays.

we entitled to?' and 'What are we trying to achieve?' were posed. To these questions there needs to be added a third, 'What can we actually prove?' The problem facing the person preparing the claim is frequently one of demonstrating what actually happened. If there are no records or witnesses to explain the delays which occurred, self-evidently, the draftsman of the claim will face a steep uphill struggle.

The cases where there is simply no evidence at all to support a claim are rare. Much more common are those where the records or witnesses exist to explain part of the story but not all of it. Unfortunately it is frequently these cases where the writer of the claim succumbs to the temptation to obscure the gaps in the evidence by the production of a claim in which the reader is confronted by an impenetrable mass of information. It may well be that in some of the cases where the court has criticised the method of presentation of a particular claim (and the *Bernhard's Rugby* case provides a good example) the claim was put in a particular way for precisely this reason.

The following checklist may therefore provide a useful starting point

- What were we obliged to do under the terms of the contract and by what date?
- When were the works actually completed – is this later than the date provided in the contract?
- Can we account for the matters which caused the works to be delayed – why have they occurred and how do we prove this? What matters give rise to an extension of time under the contract?

- What does the contract require in terms of notices and did we comply with this machinery?

Each of these questions needs a little further consideration. It is worth noting that the starting point is that the claimant should prove an entitlement to time additional to that prescribed by the contract as opposed to taking the date when the works were actually completed and attempting to fill in the gap.

The timing obligations imposed by the contract are considered in Section 2.2 of Chapter 2. The obligations governing completion of the works will be those provided in the contract. An entitlement to set aside the timing obligations provided by the contract will seldom occur in modern standard form contracts which feature extension of time provisions. As will be clear from Section 2.2, the cases where the court has set time at large are, in the main, those pre-dating the sort of sophisticated extension of time provisions found, for example in the JCT family of contracts, or else arise from tailor-made contracts where, for their own reasons the parties have inserted limited rights to extensions of time and these have proved inadequate for dealing with the delays which occurred. The point is considered in a different context in Chapter 4, but it is worth looking at the two leading cases on this issue.

Peak Construction (Liverpool) Ltd v. McKinney Foundations Ltd (1970) 1 BLR 114

The contract provided that time should be of the essence. Lord Justice Salmon's judgment does not address the specific question of when time might be rendered at large. Instead, he firstly considered the proper approach to analysis of delays and their cause and secondly determined that it was impossible in the circumstances to lay the whole of the delays at the door of the defendant. Hence he remitted the matter to the trial judge for reconsideration of the assessment of responsibility for delays.

McAlpine Humberoak v. MacDermott International (1992) 58 BLR 1

Again the contract provided that time should be of the essence. At trial, Judge John Davies held that the result of the first issue of varied drawings was (in effect) to frustrate the contract. In a contract to execute fixed amounts of work for a fixed sum within a fixed time, the consequence of the issue of these drawings was that the fixed time (and for that matter the price) went 'out of the window'. The Court of Appeal damningly referred to this as a 'novel proposition', rejecting the idea that in a fixed price contract with a firm delivery date an act by the employer which caused the contractor to miss the delivery date by a day entitled the contractor to re-cast the whole basis upon which both time and price were to be calculated. Although Lord Justice Lloyd accepted that contracts would exist where time was at large (as he said 'whatever that may mean') he stopped short of saying that this was such a contract.

Accordingly, claims which are based on the premise that the contractual machinery for granting extensions of time should be set aside and replaced by a general obligation to complete within a reasonable time should be treated with caution. Brian Eggleston notes:

> 'The phrase "time at large" is much loved by contractors. It has about it the ring of plenty; the suggestion that the contractor has as much time as he wants to finish the works. This is not what it means.'

Guidelines

Consideration of the principal standard forms suggests that certain guidelines can be drawn:

- In contracts which provide for extensions by reason of both defaults on the part of the employer and matters beyond the control of the parties, circumstances in which the extension of time mechanism can be said to have broken down will be uncommon.
- The only commonly encountered exception to this will be where it can be shown that the cause of the delay is an act of prevention or hindrance by the employer. It is settled law that if the contractor can show that completion has been prevented by the employer, the principle that the party in default should not rely upon his own breach should disqualify him from deducting liquidated and ascertained damages.[5.4] In the cases where there is

95

no, or no operable, clause for extending time this is likely to set time at large. In cases where there is an extension of time clause, the cases are silent as to whether this actually places time at large or whether, as seems more likely, the employer is disqualified from claiming liquidated and ascertained damages but the contractor is entitled to claim extensions of time in respect of the additional time spent consequent upon the employer's act of interference.[5.5]

- Where, however, the contract has been drafted or amended such that an event can occur which is beyond the contractor's control and for which the contract does not grant of extension of time, the contractor may be entitled to argue that he should be granted a reasonable time in which to complete the works (see *Scott Lithgow* v. *Secretary of State for Defence* 45 BLR 1).
- The distinction therefore needs to be drawn between contracts where the risks have been placed upon one of the parties and those where the contract simply fails adequately to deal with all the consequences of events causing delays. An example of the former is the contract in *Kitsons* v. *Matthew Hall* (1989) 47 BLR 82, which is considered above[5.6]. The latter is the position in the cases such as *Holme* v. *Guppy* (1838) M & W 387 where the extension of time machinery, if it existed at all, did not deal with circumstances where the delay on the part of the contractor had been caused or contributed to by the party complaining of it.
- The obligation to complete the works within a reasonable time will, however, arise in circumstances where the parties have failed to reach agreement as to the time for completion of the works.[5.7]

Therefore except in unusual circumstances the completion date provided in the contract will regulate the parties' obligations to complete the works. That date will obviously be varied by extensions of time actually granted.

The actual date of completion will be that measured in accordance with Section 4.2 above. While the position shown by the cases is unsatisfactory, good sense suggests that the date to be used is that when the works are complete or, to borrow Judge Newey's expression in *Emson* v. *EME*, 'complete but for *de minimis* items' – whatever this means in any given situation.

Identifying delaying events

The next question is whether the delaying events can be identified. If they can, is it possible to attribute these delays to any particular causes? If so, how is this to be proved? Are there records whether by way of diaries, correspondence or other notes, and are there still witnesses who will account for what delayed the works? What facts are going to be relied upon? These are all easier questions to ask than to answer, and the matter of proof is one frequently overlooked by potential litigants in their enthusiasm to formulate arguments. Keith Pickavance comments

'Poor quality of project documentation leads to poor factual evidence and presents serious difficulties in identifying the rights of the parties.'

He also refers to the NEDO report *Achieving Quality on Building Sites* as identifying that poor record keeping is a sign of poor management which in turn leads to a higher incidence of disputed claims. The point can be tested by reverting to the example shown in Fig. 5.2.

Hence, taking the matters set out in Fig. 5.2 and concentrating on the major assertion, the contractor's claim in respect of variations to the works and the delays which they have caused, the following points are clear.

- The existence of the instructions for additional work is not disputed. Nor is there any argument that this is work in respect of which the contractor is entitled to be paid.
- However, while it may be said with justification that these variations must have led to delays, there is no letter or minute of a meeting or diary entry which shows what any of those delays might be. Although the date of the instructions can be determined, as with the addition of dimmer switches, there is nothing to say what, if anything the consequence of this might be.
- Still less is there anything which enables the contractor to say how the delay caused by one matter may have led to delays in other respects. At best limited examples can be identified; for example with the substitution of tiles for vinyl floor covering in the bathroom, the order date for the tiles can be shown, as can the

date of delivery. The additional time compared against the time for purchase of proprietary vinyl floor covering can be proved, as can the additional time needed by the decorating sub-contractor whose men have produced daywork sheets for the additional two days work required. This in turn delayed the remainder of the finishing trades in the bathroom. However here the trail goes cold because it is impossible to determine whether this had any consequences upon the following works or upon any works to be undertaken at the same time.

The employer's assertion that there were delays in delivering the kitchen units is simpler. He can point to a date when they ought to have been delivered (provided in the programme); he can also point to the date when they were actually delivered (the architect's diary) and he can point to the effects of this on second fix electrics and decoration because of the architect's letter recording this.

Marshalling the evidence

The need for records to substantiate assertions has been touched on in Chapter 3. The above example shows that that without some form of evidence even the most plausible argument, in this case that the variations caused delays, will be difficult to sustain. So, what form can that evidence take? At this stage, we are concerned to assemble the possible material which might be used to support the claim. We are therefore not concerned with the question of whether a particular source of evidence might fall foul of the rules governing admissibility. The basic premise is that any material which goes to prove or disprove the existence of a particular fact or facts may potentially be used to support the claim.

Hence the following (while not an exhaustive list) may assist:

- the contract documents, tender, specification and drawings
- pre-tender correspondence and correspondence written during the negotiations leading to the conclusion of the contract
- correspondence between the parties to the contract, with members of the professional team, with sub-contractors and sub-consultants, with statutory undertakers and with third parties such as potential tenants or other occupiers or users of the works

including internal documents such as notes between various members of the contractor's team or reports prepared to comment upon particular aspects of the works[5.8]

- minutes of meetings, as above
- programmes for the whole or part of the works, whether intended to be 'a contract programme' or simply a document used in helping to plan and organise the works
- planning tools, including any software programmes used in planning the works
- record and transmission sheets showing, for example, the sending of a particular drawing issue sheet from one party to another, or recording the occurrence of a particular event
- diaries, whether 'official' site diaries or simple appointment diaries kept by those involved in the project
- notes of conversations and telephone calls
- photographs, videos, tape recordings and computer records
- certificates, statements or instructions issued by members of the professional team
- the statements of the personnel involved in the project.

A good guideline is that the nearer the particular piece of evidence comes to providing a contemporary first hand record of a particular event the more weight it will carry. At one end of the scale there will be a video recording (complete with date stamp) of the particular event and at the other will be a statement by someone who did not actually witness an incident himself but was subsequently told about it by another person.

Charles Dickens' Mr Bumble said 'what I wants is facts'. A fact is something irrefutable. A claim which derives entirely from facts which allow the observer (whether the opposing party, a supervising officer, a judge, arbitrator or adjudicator) to determine the whole story will almost always prevail over one which comprises snippets of information or hearsay, opinion or argument. However, the nature of the construction industry means that in most instances it is impossible easily to assemble a comprehensive account of the history of the works, and hence the importance of marshalling the available information at an early stage. At worst this will enable a party to conclude that, whatever he may believe, the facts do not support that view, or that the information required to substantiate it does not exist.

Preliminary review

Unfortunately the task of document assembly and ordering, if carried out at all, is often carried out by the most junior member of the contractor's team or by an equally lowly person within the office of his solicitors or consultants. Although it is a counsel of perfection, an essential part of this operation is for some kind of overview to be taken by an appropriate person with some knowledge of the job as to what the documents, once properly ordered, actually show.[5.9] The suggestion has been made that to ensure objectivity this should be undertaken by someone unconnected with the works. In most cases this will be impossible because it is impractical in terms of costs and the availability of senior personnel with sufficient time to devote to this task away from their other responsibilities. However it is self-evident that on this review an attempt should be made to be as objective as possible.

As well as enabling a preliminary view to be taken of the strength of the factual material, the review should also enable an opinion to be formed of the claim's weaknesses and of the areas where evidence is lacking. Three important notes of caution need to be sounded.

(1) English law operates a 'cards on the table' approach to documents. In cases commenced prior to the implementation of the Woolf Report, the obligation of the parties is to disclose *all* documents relevant to matters in issue in the litigation. While the new Civil Procedure Rules, which give effect to Woolf reforms, have the effect of simplifying the process by which this is done, the basic principle remains – the obligation is to produce all those documents, both 'pro' and 'anti' which relate to the matters in issue. The existence of damaging documents is something best identified at an early stage. Attempts to suppress such documents are generally futile – the stock-in-trade of solicitors experienced in construction disputes is to sniff out the absence of potentially relevant classes of document which may have been 'overlooked' by the disclosing party. Judges and arbitrators take a dim view of attempts to mislead them by the non-disclosure of relevant documents and both have wide powers to order draconian sanctions. Moreover, most sets of rules for the conduct of adjudication provide

100

that, where the adjudicator concludes that a party has not revealed the existence of documents which damage his position, the adjudicator can draw an adverse inference from this.

(2) The fact that such documents may be 'private and confidential' is not a reason not to disclose them. The party to whom the documents are disclosed is under an obligation, whether to the court or to the arbitrator, not to use the documents for any purpose unconnected to the proceedings. 'Privilege' – the right to withhold production of certain documents is a narrow right generally extending only to documents brought into being for the purpose of giving or taking legal advice, or which have been created in contemplation of litigation *and* where the dominant purpose of their creation is that contemplated litigation.

(3) The use of the expression 'without prejudice' will not assist if the document has come into being other than in a genuine attempt to achieve or negotiate a settlement of the whole or part of the dispute.

The next issue is whether these matters can be fitted within a particular category identified under the contract as giving rise to an entitlement to more time. Alternatively, is it possible to point to particular entitlements or rights provided by the contract which have not been met. This is considered in the next section.

5.4 The legal framework

Having identified the necessary facts, it is necessary to consider the legal framework within which the claim is made. Does this claim fit within any specific entitlement provided by the contract? For example, do the factors which are said to have caused delay correspond to any of the matters which are listed as Relevant Events under clause 25 of JCT 98, are listed in the appendix to the contract, or are affected by any implied terms (discussed in Chapter 3).

The alternative to claims brought pursuant to specific rights provided by the contract is the claim that a particular matter constitutes a breach of a right or obligation provided by the contract. Such claims are frequently but not invariably expressed as alternatives to one another. For example, a contractor may be able to point to the late provision of access to a part of the works as entitling him under the contract to an extension of time pursuant to clause 25.4.12 of JCT 98. He may also claim that under the contract he is entitled to be granted access to the works and that the employer's failure to grant access to the works constitutes a breach of contract. The key, self-evidently, is to point either to a specific right provided by the contract or to a right provided by common law. The mere fact that the opposing party can be shown to have behaved badly – or even unscrupulously – will not, without more, give the claimant a right of action.

The mistake is to believe that the absence of an identifiable express term can be made good either by the allegation of a convenient implied term or by claiming the breach of some right alleged to make good the deficiencies of the contractual claim. It is settled law that a claim for direct loss and expense and a claim for damages will be quantified in the same way[5.10] – one cannot get both. It is suggested that this principle will apply equally to the calculation of claims for additional time. Where the contract between the parties regulates the entitlement to further time, if the claimant cannot claim more time pursuant to an express contractual right, it is unlikely that he will be able to re-format his claim in terms of breach of contract – the two are necessarily opposite sides of the same coin.

This is frequently important in instances where the contract provides that the loss or delay arising from specific events will be allocated in a particular way. Examples are provided by the 'no fault' relevant events within clause 25 of JCT 98, namely

25.4.1 – *force majeure*
25.4.2 – exceptionally adverse weather conditions
25.4.3 – loss or damage occasioned by one or more of the Specified Perils
25.4.4 – civil commotion
25.4.9 – statutory regulation of labour or materials imposed after the contract was entered into
25.4.10 – the contractor's inability due to reasons beyond his control to secure labour or materials
25.4.11 – execution of works by statutory undertakers or local authorities
25.4.13 – delay due to change in statutory requirements
25.4.15 – the use or threat of terrorism.

Each of these matters will, if proved, entitle the contractor to further time and to be excused from liquidated and ascertained damages.

However, they are not also included in clause 26 and so the contract does not provide the contractor with the right to claim direct loss and expense in consequence. Unless clause 26 is modified this is how the parties have elected to deal with these risks. As Judge Edgar Fay put it in *Henry Boot Construction Ltd* v. *Central Lancashire Development Corporation* (1980) 15 BLR 1

> 'There are cases where the loss should be shared, and there are cases where it should be wholly borne by the employer. There are also cases which do not fall within either of these conditions and which are the fault of the contractor. But in the cases where the fault is not that of the contractor, the scheme [of the contract] is that in certain cases the loss is to be shared; the loss lies where it falls.'

For present purposes the distinction between claims under the contract and claims for breach of contract is important because under the terms of the Limitation Act 1980 section 5 provides that a claim brought pursuant to a contract executed under hand runs for six years from the date when the cause of action *accrues* (emphasis added) and section 8 provides that where the contract is executed as a deed the period of limitation will be 12 years. Hence, the limitation period for a claim brought under the contract will be governed by the contract whereas the limitation period for a claim brought in respect of a claim for breach will run from the date of the breach. In the case of major works, these dates will often be years apart and hence there will be circumstances in which a claim brought under the contract may be time barred whereas a claim for breach may still be capable of being brought within the limitation period.

Disruption and acceleration

Disruption

There is, however, a further issue which needs to be addressed – disruption. Commonly claim documents use the formula 'as a result of the matters identified in this claim the works were delayed and/

or disrupted', and these expressions are commonly used inter-changeably. This is unfortunate since they really refer to different situations. Delay, self-evidently, is the situation where the works take longer than originally intended. Disruption, by contrast does not refer to the timing of the works but to the situation where the works were rendered more difficult by some act of hindrance or prevention on the part of the employer. Allied to claims that the works were disrupted it is common to find the suggestion that the works were accelerated – that is that the employer required that they should be executed more quickly, either to achieve an earlier completion date or to maintain an existing completion date which had slipped as a result of other factors.

Comparatively little has been written about disruption.[5.11] It is acknowledged that disruption claims are difficult to prove. To substantiate a claim for disruption the claimant will necessarily have to demonstrate

- that the works were planned in a particular way, and
- that specific events rendered this unfeasible,
- with the consequence that additional resources had to be deployed in order to maintain progress.

In contrast to delay, where the variable which is affected by the event is time-related, here the variable is resource led.

However, the following difficulties will often occur:

- The allegation that in fact the need for additional resources was not due to any default by the employer but because the contractor himself had under-resourced his original bid, and thus the additional resources deployed were needed to make good the contractor's own default.[5.12]
- The need to link the additional resources to the act of prevention or hindrance will usually be met with the argument that the contractor's problems arose from a failure to achieve projected output levels from his own workforce.

Both are difficult arguments to disprove without detailed investigation of the underlying factual matters, including particularly the tender breakdowns for labour productivity. Self-evidently it will often be difficult for contractors to prove that their problems are attributable to an act of prevention as opposed to being a consequence of under-bidding the original tender. Put another way the burden will rest with the contractor to show

- that he could have undertaken the contract
- that his original price and output levels were realistic and
- that they would have been achieved but for identifiable events.

The practical difficulties with proving disruption claims should not detract from the fact that delay and disruption are closely related and it is common that a contract which has been delayed will also have been disrupted and that, in an attempt to minimise the effect of the delays, the contractor has taken it upon himself to increase the resources which he has deployed – hence the interchangeable use of the terms. This interrelationship is shown on Fig. 5.4. This has attempted to plot the effect of particular events in terms of both time and resource.

Event	Delay	Disruption
Failure to give access	✓	x
Failure to release whole or part of works in accordance with programme	✓	✓
Failure to provide information	✓	✓
Instruction requiring variation	✓	✓
Instruction altering sequence	x	✓
Instruction requiring acceleration	x	✓

Fig. 5.4   Delay and disruption compared.

Acceleration

Acceleration is closely related to disruption. It is only in recent years that standard form contracts have started to make allowance for the employer to issue instructions to accelerate the works. The absence from JCT 80 and JCT 81 of a specific provision for acceleration led to these contracts being extensively amended. JCT 98 has still not addressed this issue. The absence of acceleration provisions from these contracts has led to debate as to whether works might be said to be 'constructively accelerated' – that is to say that the contractor can be compelled by circumstances to accelerate performance of the works in order to avoid delays (and their consequences) which would otherwise occur. While it is possible to see circumstances in which this would happen, and which would theoretically give rise to a claim, it is suggested that such a claim would be extremely difficult to prove.

With the increasing use of acceleration provisions within standard forms this also becomes an unnecessary argument. The absence of specific provisions from JCT 98 is frequently made good by amendments.

CHAPTER SIX
LEGAL CONSIDERATIONS

6.1 When does a dispute become 'legal'?

The previous chapter dealt with putting tackle into order and the considerations involved. There is no 'right time' for instructing lawyers or even claims consultants and obvious pitfalls are associated with the premature involvement of lawyers. The risk is that it may cause attitudes to polarise and make a protracted dispute more rather than less likely – to say nothing of costs. Hence this will always be something for the individual judgement of the parties. Undoubtedly, the availability of adjudication has served to allow parties to take the decision to institute some form of dispute resolution procedure more easily.

It follows from Chapters 2 and 3 that the party who has attended diligently to the planning and execution of the works will have done everything he can to avoid this situation arising. However, it is unrealistic to suggest that this will mark the universal cure for confrontation. Disputes which cannot be settled will still arise, not least because, as we have seen, there will be instances where a genuine difference occurs where the parties hold sincere but diametrically opposed opinions. There will also be occasions where one of the parties takes a particular stance in order to improve its commercial bargaining position. In this situation litigation, arbitration, mediation and adjudication in their various forms provide simply another set of weapons in the commercial arsenal. Again, the availability of adjudication clauses make this a less destructive course of action.

One should not demonise construction disputes. The key is to understand the place which disputes and the available mechanisms for their resolution occupy in the construction process.

The condemnation of the excessive occurrence of litigation as a means to settle disputes has largely concentrated on four basic premises:

(1) Confrontation has been caused by onerous and one-sided amendments to standard forms, often drafted by lawyers or surveyors with the object of improving their client's position at the expense of fairness.

(2) This has led to a culture in which litigation has become the only way in which a party can actually protect his position because the contract conditions promote conflict.

(3) Litigation itself is a blunt instrument which seldom produces a result that fulfils the commercial aspirations of the parties, but instead is slow, labour intensive and stressful.[6.1]

(4) The consequence of this has been to mould an industry in which insufficient attention is given to construction itself, and excessive regard is had to creating or defending the arguments inevitable in such a climate.[6.2]

As an aside, these comments are not affected by the existence of adjudication which, after all, is simply intended to be a quicker way to resolve disputes. In this respect, adjudication will not mean there are fewer disputes. On the contrary, it will make it simpler and cheaper to use a form of formal dispute resolution to deal with the differences that inevitably arise.

All of the above comments are largely justified. However, they fail to take account of one simple truth. The problems which have undoubtedly bedevilled the industry owe a great deal to a failure by the parties either properly to understand the nature of their obligations or adequately to plan their implementation. Inevitably this has caused and will cause disputes when the parties come to construct the works.

This has been demonstrated time and again in recent years. In a badly planned, inadequately understood and hence inappropriately resourced and poorly performed contract, in which insufficient attention has been paid to any prevailing contractual requirements, a claim is almost inevitable. It hardly needs saying that the likelihood of that claim having any merit is not enormous.

Sadly, it is often precisely these claims which are hardest to resolve because the perceptions of the parties will frequently be so different. The result is generally one which will leave both parties profoundly dissatisfied; if the claimant recovers only a fraction of what he is seeking he will feel that he has not got his due while even if the respondent has substantially defeated the claim, the outlay of irrecoverable costs and management time may leave him feeling that the system has failed him.

It is relevant to consider the changed climate brought about by adjudication. The relative ease with which adjudication proceedings can be brought means that disputes can be dealt with as part of the project. It does not require the parties to switch into 'dispute mode' and there is a better chance of preserving an amicable working relationship.

6.2   *Claims as negotiating tools*

Once it becomes apparent that the dispute will not settle without at least a preliminary skirmish, it is worth considering what will have to be proved in order to win in adjudication, arbitration or in court. Experience suggests that the cases which tend to settle are those where the parties have taken time to determine what they will have to prove and whether they can actually achieve it; generally these settle on the best terms and most painlessly. In the case of adjudication where the compressed time scale means that more cases 'go the distance', the prospects of success are increased. For the purposes of this chapter it has been assumed that the same standard of proof will be required in adjudication as in litigation or arbitration. This view is considered in a little more detail in Chapter 9.

Ultimately, the question is – do we have the records, correspondence, witnesses and documents to prove our case on the balance of probabilities?

What does this mean? On the simplest view it is that on balance a particular version of events is likely to be true. In delay claims this translates into the proposition that one party's explanation of the delay is more likely to be true than the other's.

Again, consideration of the questions in Section 5.3 leads to the conclusion that if these points can be successfully addressed, there will be good prospects of proving the claim. If, on the other hand, a dispassionate analysis leads to the conclusion that there are substantial problems with proving these matters, it is sensible for the party in question to re-appraise his objectives in a particular claim. In short, while there is no reason why a party should not commence proceedings in the knowledge that he may face great difficulties in making his contentions stick, he should appreciate that this is a risky policy. What he is doing is litigating as a means to engineer a negotiating position from which he can achieve a better result than he would have done otherwise. With adjudication, this approach is especially risky because it is less likely that the matter will settle during the brief time scale of the adjudication process.

Tactically, this approach comprises a game of bluff. The secret is in putting a case at its highest in the hope that this will persuade an opponent to the view that, even if he believes that the case against him will ultimately fail, there is enough in it to convince him to put a price on the perceived risk which he faces. While the broader tactical implications of prosecuting and defending delay claims are considered below in rather more detail, experience teaches two important lessons:

- Even the best claims are usually a mixed bag comprising matters which can and cannot be proved. The prudent claimant is the one who has taken time to analyse the claim's overall prospects of success at the outset. The prudent respondent carefully assesses what is being said against him with a view to evaluating the real risk which he faces from each particular claim.
- Construction litigation, adjudication and arbitration in all their various forms are structured in a way which favours the party who pays the most attention to getting his case right from the outset. While of course it is possible to correct mistakes by amendment, this provides a tactical advantage to the other party.

Common sense alone suggests that if someone is to go to the trouble and expense of some form of formal dispute resolution, he should put his best foot forward. If nothing else, against the acknowledgement that most claims will have an Achilles heel of some sort, by putting the claim in its best light the chance of creating in an opponent the belief that settlement is the best course to take is maximised. In the sort of case which really is started simply as a negotiating ploy, to do otherwise is to risk being sucked into litigation with little prospect of success and no retreat other than a potentially humiliating and expensive climb-down.

This is all very well in the abstract. The question of how the various matters considered in this book so far translate into preparation of claims has considerably vexed the construction industry, lawyers and consultants alike.

6.3   *Showing cause and effect*

At the heart of the debate is the principle enunciated above – in order to prove a delay claim a party must show that, on the balance of probabilities, the delays complained of arose by reason of matters which entitle him to an extension of time. That has been refined into the premise that he must prove not only the events themselves but that they actually caused the delays. In other words, cause and effect.

This allows us instantly to say what a claim should not be. That has been described as the 'rolled up' or composite claim. This expression has been used in a number of contexts, and a good deal of time has been spent by lawyers arguing over whether particular claims were or were not rolled up. However, the terminology itself is unimportant and loses sight of the basic principle. When a claim fails adequately to identify both the events which give rise to the entitlement and the inter-relationship between those events and the delays themselves, this will not discharge the burden of proof.

Wharf Properties

To the extent that it is necessary to prove this point, one only needs to look at the now notorious decision of the Privy Council, *Wharf Properties v. Eric Cumine Associates* (1991) 52 BLR 1.[6.3] Like many milestone cases in recent years, this is a decision which has been held out as providing authority for a great many propositions which cannot really be distilled from the judgments delivered in either the Hong Kong Court of Appeal or in the Privy Council.[6.4] While this is a case which serves to place earlier cases in their proper context, the correct view is that it merely states the proper state of the law then and now. However for that reason alone it is an important case and hence merits careful consideration. What it undoubtedly does is to show that in years to come claimants will need to be more diligent in the way in which they approach the question of causation.

The facts

The dispute concerned the construction of a major residential and commercial development called Harbour City in Kowloon, Hong

Kong. The plaintiffs were developers who engaged the defendants, a well known firm of Hong Kong architects. The works were substantially delayed and Wharf commenced proceedings against not only the architects, ECA, but originally also against the main contractor, the consulting engineers, and fifteen others, including most of the principal sub-contractors. The claims against all bar ECA were settled and the proceedings were reconstituted as a claim against the architects alone.

In essence, Wharf alleged that ECA had caused the works to be delayed because they had failed properly to manage, control, co-ordinate, supervise, or administer the works of the main contractor and sub-contractors. This was expressed as both a failure of supervision and also a failure properly or timeously to supply necessary information. They had also issued excessive numbers of variations and instructions, which had caused the works to be further delayed. The complaint was aimed both at the number and the timing of the variations. Consequently, Wharf had become liable to pay nearly HK$318 m to the contractors, including the sums paid to settle the proceedings, and also alleged a loss of rental income consequent upon the delays amounting to nearly HK$200 m.

The pleadings

These allegations were contained in a statement of claim which even after settlement of all bar the claims against ECA ran to 155 pages, supported by schedules which comprised over 300 further pages. Somewhat unusually, the pleading was divided into sections (an unsatisfactory expression which caused confusion with separate portions of the works) rather than numbered paragraphs. Both in the Hong Kong Court of Appeal and in the Privy Council comments were made that, while the judges were conscious of the difficulties facing any party attempting to articulate such a complex claim, in this instance the need to cross refer to two or more separate documents in order to follow through any particular allegation was immensely confusing.

These difficulties would have been awkward but manageable had it been possible to identify exactly how the facts and the delays they were alleged to have caused actually interlinked. As Mr Justice of Appeal Power succinctly put it: 'Nowhere however does the pleading indicate which periods of delay are due to the alleged mismanagement of ECA'. Indeed at first instance Mr Justice Mortimer had summed up the case pleaded by Wharf as alleging that while some of the delays were explicable and forgivable, the remainder collectively had been the cause of delays and were due to the default of ECA. There was no attempt to identify individual delays which caused particular delays.

The exact nature of the problem is minutely analysed by Lord Oliver. He started by taking the allegation which appeared in section 6 of the pleading, that particular periods of delay had been caused in relation to the excavation and groundworks, and noted that this could not be connected to any particular allegation against the architect, but that Wharf had made no more than a generalised complaint of negligence and breach of contract. He went on to state tersely that the promise that these delays would be explained in later sections was not made good. While a variety of failures were advanced none of these actually linked in to any of the periods of delay. The best that could be said was that ECA had failed to perform 'timeously'. Lord Oliver noted that:

> 'Since the pleading nowhere states what "timeous" performance would have been this amounts to no more than saying that ECA were late by some unspecified margin in doing what it was their contractual obligation to do at some unspecified point in time.'

Unsurprisingly, ECA delivered a substantial request for further and better particulars, running to 357 separate requests. Lord Oliver drew attention to two of these: request 34 which sought details of the allegedly excessive variations and asked for a Scott Schedule detailing their effect, and request 44 which asked Wharf to particularise the delays which these matters had caused. Wharf failed to serve the particulars and after an application to the court, ECA obtained an order that they be provided. Of the contents of the document which was eventually served, Lord Oliver dismissively observed: 'they did not in fact comply with the order at all, nor can they be said even to merit the description of "particulars".'[6.5]

What Wharf said was no more than that the number of variations was excessive and taken cumulatively they had caused delay. With an honesty which bordered on the self-destructive they went on to say that at trial it would be necessary to look at each and every one of them in order to see which were justified and which were excessive. They admitted that they were unable to attribute individual causes to the delays. The single glimmer of hope offered to the reader was the promise at some future unspecified date following discovery to provide a critical path network showing the delays caused. Even this, it will be noted, does no more than offer to trace the path of the delays rather than showing how they had actually been caused.

The summons to strike out

ECA issued a summons seeking to strike out the statement of claim on the grounds that it disclosed no reasonable cause of action, alternatively that it was an abuse of the process of the court, or alternatively that it should be struck out unless proper particulars were served. Wharf issued an application of their own seeking an order that ECA should provide discovery of all documents which related to the requests. At first instance, Mr Justice Mortimer refused the application to strike out and made the order for discovery.

ECA appealed and the Hong Kong Court of Appeal struck out the statement of claim (sparing only a claim in respect of a performance bond and a claim resulting from additional excavation work, neither of which were affected by the problems which beset the remainder of the pleading.) Mr Justice of Appeal Power held that the statement of claim disclosed no reasonable cause of action. In a short concurring judgment, Mr Justice of Appeal Penlington gave the additional reason that the statement of claim was an abuse of the process.

The Privy Council

Wharf appealed to the Privy Council. Lord Oliver declined to follow the reasoning of the Court of Appeal, concluding that if the pleading alleged a contract and then pleaded a breach, that did disclose a cause of action which could entitle the plaintiff to at least nominal damages. He accepted the theoretical possibility that Wharf might, if they proved each and every fact upon which they relied, be able to make good their contentions. The enormous evidential difficulties which this posed did not of itself justify the court in striking out the action.

However, he then found that, as pleaded, the statement of claim was hopelessly embarrassing to ECA.

> 'The failure even to attempt to specify any discernible nexus between the wrong alleged and the consequent delay provides . . . "no agenda" for the trial.'

Looked at against a background of a claim which sought colossal sums and had dragged on for five years in what amounted in his view to a concerted attempt by Wharf to evade and prevaricate, there was ample material to justify Mr Justice of Appeal Penlington's view that the claim was clearly an abuse of the process. He robustly declined Wharf's gallows plea for one last chance to put their case in order.

Reactions to *Wharf*

The case has been seized upon as setting a benchmark for the way in which claimants should go about proving their case. It has been widely cited by respondents alleging that a claim which has been advanced is fatally flawed. These contentions have taken three main forms.

Failure to link cause and effect

The most significant of these is to allege that the claim fails properly to link cause and effect, that the claim does not demonstrate what Lord Oliver described as the essential nexus between the events causing the delays and the delays themselves. This point is well made although, as will be seen below, it does not of itself provide a respondent with an automatic method of defeating a claim. As Lord Oliver and Mr Justice of Appeal Penlington both held, the case advanced by Wharf did disclose a cause of action, which might have succeeded, even if that seemed highly implausible; what proved fatal for Wharf was the degree to which the way their case was stated was prejudicial to ECA.

Accordingly, the suggestion that a claimant must adopt a particular approach in order to escape the implications of the decision is mistaken. At its most extreme, this argument manifests itself in the contention that the claimant will fail unless he pleads his case by way of a critical path analysis, and that in the absence of this a claim will be doomed. This point is considered in depth in the next chapter. However, it is sufficient for present purposes to say that critical path analysis, itself a widely abused and oft-misunderstood concept, may not be the answer in every case, and, indeed, will sometimes be completely inappropriate. Provided that the claimant makes the connection between cause and effect, the method adopted will not be important.

'Rolled up' claims

Equally mistaken is the argument that this marks the death knell for the 'rolled up' claim in any form, and that in consequence the earlier decisions in *J Crosby and Sons Ltd* v. *Portland UDC* (1967) 5 BLR 121 and *London Borough of Merton* v. *Stanley Hugh Leach Ltd* (1985) 32 BLR 51, were no longer to be regarded as being good law.[6.6] This is a surprising observation when one considers that both cases were referred to the Privy Council, who decided that neither were really germane to the issues before them. It is instructive to consider why this view was taken.

In *Crosby* the dispute related to the laying of a water main under a contract incorporating the ICE 4th Edition. An arbitration took place following which the parties and the arbitrator referred a series of 29 questions to the court.[6.7] Of these the only issue which really concerns us is the last which dealt with the proper method for assessing the contractor's claim for extra costs. Mr Justice Donaldson identified nine separate matters which had caused delay to the contractor. The arbitrator had found as a fact that these had caused a total of 31 weeks delay to the contractor, and had stated:

> 'As each matter occurred its consequences were added to the cumulative consequences of the events which had preceded it. The delay and disorganisation which ultimately resulted was cumulative and attributable to the combined effect of these matters. It is therefore impractical if not impossible to assess the additional expense caused by delay and disorganisation due to any one of these matters.'

What is significant is that the arbitrator had felt able to find that the delays had been caused by the various causative matters and thereafter had been concerned solely with determining the proper way in which to assess their financial consequences. He proposed the making of a lump sum (or rolled up) award. The respondents argued that the contract contained an elaborate machinery for assessing and adjusting the sums payable under the contract and that hence the claimants should prove the sums separately attributable to the individual heads of claim.

The judge conceded that the question was not easy, but adopted a practical solution:

> 'I can see no reason why [the arbitrator] should not recognise the reality of the situation and make individual awards in respect of those parts of individual items of the claim which can be dealt with in isolation and a supplementary award in respect of the remainder of those claims as composite whole.'

This approach was strongly approved by Mr Justice Vinelott in *Leach* who added the helpful comment that it was:

> 'implicit in the judgment of Mr Justice Donaldson first that a rolled up award can only be made in a case where the loss and expense attributable to each head of claim cannot in reality be separated and secondly that a rolled up award can only be made where, apart from that practical impossibility, the conditions which have to be satisfied before an award can be made have been satisfied in relation to each head of claim.'

In other words, the question of whether it is appropriate to make a rolled up award, and by implication to advance a rolled up claim for loss and expense, only arises when the question of the delays and their causes has been disposed of.

The obvious practical lesson which this teaches is that, while claims for loss and expense can be made in a rolled up form, the cases do not provide an excuse for laziness, and where the claim can be broken down to allocate specific items of loss and expense to particular causes this should be done. The reverse side of this is that it would be possible in theory for a respondent to say that a claimant could have carried out this sort of exercise, although in the main this will not mark a sensible course to take, if only because it comes perilously close to asking the respondent to prove the case against him. However, and as will be demonstrated, it can provide a useful way for a respondent to attack the causal link between the delays and the events which caused them and the financial consequences which are said to flow from them.

It will therefore come as no surprise that in *Wharf* Lord Oliver described himself as 'wholly unpersuaded' that either *Crosby* or *Leach* had any bearing upon the matters before him. He noted that ECA were not concerned with the financial consequences of an award but with the preceding issue of proving the delays themselves.

Obligations of the professionals

The third and final issue which has been widely argued in the wake of *Wharf* is one which is generally limited to claims against consultants, although in some instances it may have a wider application. That is the suggestion that it is wrong to advance a claim that a member of the professional team failed to procure completion of the works by the date stated in the building contract. Unsurprisingly, this argument has gained particular popularity among professional indemnity insurers. Unfortunately, the decision, while providing some useful insights into this argument, does not provide any generally applicable principle for the simple reason that it was unnecessary in the context of the issues before the court to decide this question.

In any event the point can be dealt with quickly. What has been said in the wake of *Wharf* is that, in the absence of any term in the contract between the relevant professional and his client – the employer or the contractor (depending upon the contractual regime which is being used) – to the effect that the professional will so carry out his duties that the works will be completed by the completion date in the building contract, no such term can be implied and therefore the court should not fix the professional with responsibility for delays. Certainly it was the case in *Wharf* that there was no attempt made to allege that ECA themselves owed a duty to ensure that the works would be completed by any particular date and the absence of such an allegation was noted. However, to suggest that this will undermine a great many claims against allegedly negligent professionals is to over-simplify matters.

This is because, in the case of architects at least, their appointment will be on terms that they should use the skill and care of the reasonably competent architect experienced in carrying out work of that size and complexity. Under the terms of the building contract the architect will be empowered to issue instructions and to order variations. Accordingly it is entirely possible that if the appropriate degree of skill and care is not used in the exercise of these powers, delays may be caused which are the direct result of the poor performance of the architect. Applying first principles, the losses which result from this will flow naturally from the architect's breach of the terms of his appointment, or alternatively they were or should have been foreseen by him as the natural consequence of his failure.

On this basis, the case will actually serve to undermine relatively few professional negligence claims. However from the viewpoint of insurers it is important, because it makes it clear that those mounting claims against members of the professional team have to take rather more care over the way in which such claims are formulated than had sometimes hitherto been the case – where the fact that one of those involved in the project had the benefit of professional indemnity cover has almost been regarded as reason enough to found a claim.

General

Accordingly, it is difficult to draw the conclusion that the Privy Council intended that *Wharf* should provide much in the way of general guidance to those advancing delay claims. Inevitably, the Privy Council was substantially concerned with the facts of the particular dispute before it. Indeed, in almost the first paragraph of his speech Lord Oliver noted that, while Wharf had argued that the case raised questions of general importance on the way in which the courts viewed *Crosby* and *Leach*, as the arguments had progressed it had become apparent to him that the case was really concerned with a point peculiar to the particular case. If there is a wider principle to be drawn, it should be that the onus is upon the claiming party to prove his case.

From that the conclusion is obvious – not only must a claimant prove the facts which he says caused the works to be delayed, he must then show precisely how those facts caused the delays. The case is therefore much more concerned with what not to do than with any attempt to say how a claimant should go about the task in hand. Hence, other than the general principle summarised above, attempts to make the decision in *Wharf* authority for any more specific propositions are misconceived. Indeed, there is nothing very novel about the general principle which may be derived from the case. In some respects it is no more than a restatement of the obvious:

- firstly that in order to succeed the claimant must prove his case on the balance of probabilities
- secondly that in order to do this he must properly plead his case.

Cases since *Wharf*

The dangers of attempting to distil precise rules from the decision in *Wharf* are illustrated in two cases both of which have been heralded as signalling the start of the retreat from *Wharf*. In fact, it follows from the above that, while expressions of this sort are useful in terms of providing titles for articles, that is their only purpose; and in the present case, since *Wharf* scarcely covers any new ground, it is a little difficult to see what is being retreated from.

Mid Glamorgan v. Devonald Williams

The first of these is the decision of Mr Recorder Tackaberry in *Mid-Glamorgan County Council* v. *J. Devonald Williams & Partner* (1991) 29 Con LR 129, a mere seven months after *Wharf*. This was another claim by an employer against an architect. Much like *Wharf* this was a case in which the plaintiff seems to have found great difficulty in putting his case in order. However the similarities really stop there. The claim itself was of a modest size, and the range of complaints raised by the employer against the architect was considerably more limited. Although, as before, a concerted attempt was made to extract particulars of the claim and the particulars when eventually served were sketchy and generalised, it is not clear whether the defendant's concerns were directed towards the way in which the plaintiffs put their delay claim, the linkage between the delaying events and the delays said to result, the plaintiffs' monetary claims or a combination of the three. In any event, there is ample material in the extracts from the pleadings which the judge quoted to criticise all of these matters.

Unhappily, there is no clear indication in the judgment as to the precise way in which the defendant's application was framed, although as in *Wharf*, it was suggested that the pleaded case disclosed no cause of action or alternatively that it was embarrassing or prejudicial to the fair trial of the action. On this basis it may be that the defendants took the view that this was simply a case which had no realistic prospect of success and therefore, given the insuperable obstacles faced by the plaintiffs, their claim should be put out of its misery. Although this sounds flippant tactically, it would not be a surprising approach for the defendant to adopt. Such applications are not unusual where the defendant feels that, while on a strict application of the rules there is no great likelihood of his application succeeding, it may be an effective method of conveying to the plaintiff that this is a case where he should lower his expectations and settle.

It is interesting but ultimately unproductive to speculate. What is apparent is that the plaintiffs made it clear that the case they had advanced marked the best they could do. Taking this cue, the judge concluded that, although the plaintiff had been slow and haphazard in its approach to particularising its case, there was no discernible attempt to evade its obligations or conceal the true nature of its case. Following Lord Oliver in *Wharf* the judge concluded that the case pleaded by the plaintiff did disclose a cause of action and that it was at least possible that this case would be successful, albeit that it posed colossal evidential difficulties. Nonetheless, it was for the plaintiff to appreciate this fact and ultimately for the matter to be decided by the trial judge. In concluding that this did not merit the 'draconian' remedy of striking out the statement of claim, it follows that he impliedly held that the case before him did not fall within the 'no agenda for the trial' criticism identified in *Wharf*.

If the case is left there, it is clear that Mr Recorder Tackaberry applied the broad principle from *Wharf* and concluded that the degree of prejudice faced by the defendant did not equate with that in the earlier case and that therefore he would not be justified in exercising his discretion. However, the case has an interesting coda, which may serve to explain some of the sound and fury which *Wharf* has stimulated. Referred to *Crosby*, the Recorder quoted the passage which is set out on page 117 above and made the observation

'I remind myself again that the learned judge was hearing a case stated on the question of whether an arbitrator was entitled to make a composite award in respect of the time consequences of a number of disparate events. Subject to certain conditions, his decision was that as a matter of law, the arbitrator could make such an award.'

With deference to Mr Recorder Tackaberry, this is simply not correct. It is clear from the passage quoted and the point is left beyond doubt by the immediately preceding passages in Mr Justice Donaldson's judgment in *Crosby*, that he was solely concerned with the status of a rolled up claim for loss and expense in circumstances where there was no issue about time. That point is emphatically confirmed by Lord Oliver in *Wharf* in the passage which is paraphrased above on page 118.[6.8]

On that basis, the decision in *Devonald Williams* is perhaps best treated as one which adds nothing to what was stated in *Wharf* and the principle which can be derived from that case. The observations of the judge which are quoted in the immediately preceding paragraph are not essential to the conclusion which the judge eventually reached and can best be treated as 'off the cuff' remarks to which little weight should be attached.

GMTC Tools v. Yuasa Warwick Machinery

The second step in the so-called retreat from *Wharf* has come in the decision of the Court of Appeal in *GMTC Tools and Equipment Ltd* v. *Yuasa Warwick Machinery Ltd* [1995] CILL 1010. Perhaps surprisingly the most striking feature of this case is that *Wharf* does not seem even to have been cited to the court. While the case marks a useful commentary on the way in which a party should approach the question of how best he should put his case and on the dangers of attempting to force an opponent to put his case in a particular way, there is nothing in either of the judgments which affects the broad principles of proof and causation derived from *Wharf*.

Rightly, Lord Justice Leggatt observed at the start of his judgment that the case did no credit to the legal system. By the time it reached the Court of Appeal, it had been on foot for over six years and had lurched through a series of preliminary hearings and amendments to the way in which the parties put their cases. The dispute concerned a computer operated lathe used for manufacturing metal blanks which were then milled using a separate machine to make rotary cutters. The lathe proved to be seriously defective. Over a three year period it worked only intermittently. The plaintiff was therefore unable to guarantee its output of blanks and had to buy them from elsewhere. It claimed the additional costs of this together with claims in respect of the downtime during which the lathe did not operate properly. A separate claim was made in respect of the wasted management time which this had caused. After the plaintiff was asked to provide a variety of particulars of the way in which the losses said to flow from the various breakdowns occurred, an exercise which seems to have caused tremendous difficulties, the judge directed the service of a Scott Schedule linking each separate breakdown to the consequent downtime and management costs.

After a number of attempts to put its claim in this way, the plaintiff effectively conceded that it was unable to plead its case in a way which linked the individual breakdowns to precisely calculated periods of downtime in the manufacturing process for finished cutters. Judge Potter struck out that part of the plaintiff's case, the plaintiff appealed and before the Court of Appeal it was argued on behalf of the defendant that the link between breakdown and downtime was essential to establish the necessary 'nexus' between cause and effect.[6.9]

Lord Justice Leggatt took a pragmatic view of this argument, noting that this was to miss the point so far as preparation of the claim was concerned. That was that after a particular date the lathe had operated properly and this had permitted the plaintiff to produce its full requirement of blanks. In reality, the effect of any one breakdown did not automatically lead to a period of downtime. Hence, the plaintiff's loss was measured by subtracting production during the period when breakdowns had occurred from a similar period when the lathe was working properly. He did not shy away from noting that this approach posed a series of problems of proof for the plaintiff but indicated that the suggestion that the case could only be put in the manner contended for by the defendant was plainly wrong and that it would be mistaken to force the plaintiff to adopt this approach. He summed up the position by saying that:

'No judge is entitled to require a party to establish causation and loss by a particular method, especially when that method proceeds, as happened here, on what can only be described as an imperfect understanding of the plaintiff's manufacturing process.'

He went on to say that a plaintiff should be permitted to formulate his case according to whatever method he saw fit and should not be forced into a straitjacket of the judge's choosing. This sentence was echoed by Lord Justice Simon Brown in a short concurring judgment, who also noted that this was particularly the case when, as here, the plaintiff did not seek to put its claim in this way but had adopted an altogether simpler method.

So does *GMTC* mark the start of a gradual lessening of the rigours imposed by *Wharf*? The editor of *Construction Industry Law Letter* suggested as much in terms in the editorial to the February 1995 edition. He went on to make three points:

(1)   He stated that in *Wharf* the Privy Council had struck out a claim which failed adequately to link cause and effect. This is certainly true, but, in failing to say that this was done because, on the facts of that case, the statement of claim was held to be 'embarrassing' to the defendant (i.e. the claim cannot be properly responded to due to the lack of proper detail), he risks missing the point of the decision.

(2)   He added that in some instances, parties had actually succeeded in having their opponent's cases struck out and he referred in support to *ICI* v. *Bovis* where in fact the judge had actually refused to do this but instead had merely ordered the plaintiff to serve a Scott Schedule which made sense of its contentions. This case and its implications will be considered below. In contrast he noted that in some cases, principally arbitrations, rolled up claims have been allowed. The reality of the situation is that, although *Wharf* does not act as a bar to rolled up claims, few construction lawyers will miss the opportunity to criticise a claim which fails to link cause and effect and its value is to place judges and arbitrators alike on notice that faced with a claim which fails to make this connection they will be urged to the conclusion that the opposing party faces an uphill struggle to make good his case.

(3)   He concludes by criticising the time and effort which have been expended by parties to litigation in attempting to *Wharf*

their opponents in circumstances where the only purpose which this has served has been to make more work for lawyers, and notes that seldom will the pleadings in a case actually provide the 'agenda for trial' anticipated by Lord Oliver. This is largely to evade the real point of the rules of pleading which are there to require a party to let his opponent know precisely what will be said against him at trial and to meet it as best as he is able. In *Wharf* the inadequacies of the plaintiff's case meant that the defendant had no prospect of knowing, on the basis of the plaintiff's pleadings, what was to be alleged.

However, while it is certainly apt to draw attention to the case, the reality of the matter is that *GMTC* is a case in which the court determined that the plaintiff's case was not, on the facts, appropriate to be tried along *Wharf* lines. It is frankly difficult to see how suggesting that one of these decisions is in some way qualified in its effect by the other will serve to do anything other than to cause wholly unnecessary confusion. It is equally difficult to support the contention that *Wharf* has actually created additional work for lawyers and claims consultants. The reverse would actually seem more likely. The 'fall-out' from *Wharf* has been to stress the importance of properly pleading causation, precisely in order to avoid the sort of procedural drubbing which was administered in *Wharf*.

McAlpine Humberoak v. McDermott

The dangers inherent in attempting to draw a hard and fast set of rules in relation to the proper method for assessing delays are demonstrated in *McAlpine Humberoak Ltd* v. *McDermott International Inc.* (1990) 24 Con LR 68 and (1992) 58 BLR 1. As we have already seen in Chapter 2, the case concerned the construction of a drilling platform for a North Sea Oil rig. At first instance, Judge Davies concluded that the approach adopted by the defendant's expert (which he described as 'a retrospective and dissectional recreation' of the contract by experts) in order to ascertain the exact effect of various alleged delaying events was unhelpful, artificial and ultimately of no particular use in deciding how delays had actually been caused. The Court of Appeal, by contrast, decided that the minutely detailed reconstruction of the impact of delaying events by the defendant was 'just what the case required'.

Again, it does not appear that *Wharf* was cited even though the Privy Council had rendered its decision over a year before the Court of Appeal came to give judgment. The reasons are obvious. *Wharf* was a decision on particular facts and the lessons from that case will be applicable in some instances where the court is called upon to decide whether the plaintiff's case has been proved, but is certainly not to be regarded as setting the universal standard by which all subsequent cases on causation will be settled.

ICI v. Bovis

The next case which needs to be considered is *Imperial Chemical Industries plc* v. *Bovis Construction Ltd* (1992) 32 Con LR 90, another interlocutory skirmish, in this instance arising from the refurbishment of the plaintiffs' head office and where the matter came before Judge Fox-Andrews. As a sidenote, this judgment should be model for arbitrators and lawyers trying to get to grips with this subject – it leaves nothing to the imagination and provides a most comprehensive analysis of the subject.

Unlike *McAlpine* not only was *Wharf* plainly in the contemplation of the parties but is also thoroughly analysed in Judge Fox-Andrews' judgment. The plaintiffs claimed that by reason of their breaches of contract, each of the three defendants, the management contractor, the architect and the consulting engineer, had caused the costs of the works to escalate by some £19 m. In the statement of claim no particular attempt was made to link any specific breach to any particular loss. The plaintiffs were ordered to particularise their claims in the form of a Scott Schedule. This was duly served, and judging from the extracts which are quoted in the judgment, the plaintiffs certainly attempted to provide a breakdown of the sums which they claimed. However, this was not done by means of an analysis of the delays which were attributable to particular events which could then be linked to the sums which might be said to flow from these events as had been done in *Crosby*. Instead, it was said that sums had been wrongly certified by the architect in favour of the contractor, these sums should not have been so certified, and that hence the sums were recoverable as damages from the three defendants.

126

In order to meet the obvious criticism – that if the defendants could show that part of the cost overrun arose by reason of some default on the part of the plaintiffs the claim would fail, the plaintiffs attempted to show that individual items of cost arose by reason of individual matters. Unfortunately, this attempt by the plaintiffs to particularise their case was subjected to two criticisms by the judge. The first was that a number of the individual allegations were put in the vaguest terms imaginable. As an example, the judge took the claim in respect of labour disruption to sub-contractors. In respect of a claim for some £1.4 m by the mechanical services contractor, the plaintiffs admitted that they were unable to provide any explanation for the sums incurred. The judge noted that this underlined the unsatisfactory way in which the case was put.

The second criticism was that in other instances the claims advanced simply defied common sense. The judge considered a claim in respect of a group of work packages where a total of £840 211 was claimed. The defendants had asked what happened if all bar two items, the changing of an electrical circuit and the re-positioning of a firebell, were shown to have occurred by reason of matters for which the defendants could not be blamed. To this the plaintiffs said that since the sums had actually been paid, their claim would simply fall to be divided among a smaller number of items. The judge described the proposition that a sum of over £840 000 could represent the cost of repositioning a firebell as being 'palpable nonsense'.

The judge considered the effect of *Wharf* and *Crosby* and, while making the observation that a claim should be pleaded properly, and indicating that where allegations of breach and their consequences are not adequately particularised, a striking out application may be appropriate, he did not attempt to set out any rules of universal application, indicating that in his view it was a question of degree in each case. In the present proceedings, he found that there was no question of striking out the plaintiff's case but that, given the inadequacies of the Scott Schedule, it was apt to order the plaintiffs to serve a fresh schedule, properly pleading their case.

British Airways Pension Trustees v. Sir Robert McAlpine

The penultimate case is the December 1994 decision of the Court of Appeal in *British Airways Pension Trustees Ltd* v. *Sir Robert McAlpine*

& Sons Ltd (1994) 72 BLR 26. Like *GMTC* and *Devonald* it has been suggested that this case should be read as marking the trend away from the position described in *Wharf.* However, like those cases, the reality is rather different.

The case differed from the majority of decisions on this subject because it is not a delay claim at all. Instead, it concerned defects found in an office building. Superficially, therefore the application of this case to delay claims is not instantly apparent. However, what makes this decision interesting is the way in which the parties' approach corresponded closely to that adopted in the delay cases. The defendants' complaint was that it was impossible on the basis of the particulars provided to apportion the sums claimed in respect of the various defects among the various defendants because, although the defects were identified individually, the sums which were claimed were not allocated to the defects individually. The defendants applied to strike out the statement of claim and hence the action. Judge Fox-Andrews accepted that the pleading was embarrassing to the defendants and granted the striking out order.

On appeal the Court of Appeal approached the matter both practically and robustly and in so doing demonstrated the dangers in attempting to draw generally applicable principles from this line of cases. Lord Justice Saville in the leading judgment adopted the now traditional approach of noting the purpose of pleadings, but proceeded to depart from the policy of then condemning the way in which the plaintiffs' case was put. He indicated that in the present case, since the defects were alleged in respect of works which the defendant main contractor had themselves carried out, it should not be too difficult for them to carry out an apportionment between the matters for which they could be held liable and those matters which in reality were the responsibility of sub-contractors. This may be important as a way to differentiate this case from the delay cases where the complaining party will generally be attacking the way in which delays suffered by someone else are pleaded.

As to the failure otherwise to divide the claimed damages between the various heads of loss it was held that, although this was a serious defect in the pleadings, it was one which could be cured and indeed the plaintiffs had offered particulars during the course of the hearing. It was certainly not a case which merited the draconian sanction of a striking out order. Lord Justice Saville distinguished the case from *Wharf* by noting that this was not a case where there had been any express refusal to provide particulars.

It is vital to a proper understanding of this case to appreciate that the Court of Appeal was concerned with its particular facts. It is implicit in both the judgment of Lord Justice Saville and also the short concurring judgment of Lord Justice Beldam that they were concerned to deal with the matter in hand and were not attempting to lay down any hard and fast rules. Lord Justice Saville expressly declined to offer a view as to the plaintiffs' prospects of success. If the case has any lesson at all for the future it is perhaps that an application to strike out will only succeed in the most extreme circumstances – such as those in *Wharf*, and that although this may be a worthwhile tactical ploy in order to force a party to confront the weaknesses in his case it is probably not an approach with much to commend itself from a legalistic standpoint. At the same time, it should not be forgotten that the court's observations concerning the proper purpose of pleadings and the need to particularise a claim exist irrespective of the precise nature of the case.

Bernhard's Rugby Landscapes v. Stockley Park

The recent cases show that the courts have looked at each case in turn. This is shown by the decision of Judge Humphrey Lloyd in *Bernhard's Rugby Landscapes* v. *Stockley Park Consortium* (1997) 82 BLR 39. This case is considered elsewhere in relation to a number of lessons which it provides in both claim preparation and in the management of the material upon which a party will seek to rely. It is important here because it provides a summary of both the previously decided cases and more significantly, the principles to be followed by the court in determining whether a claim had been properly particularised and by the parties in formulating that claim.

The dispute concerned the construction of a golf course by landscape contractors. After a lengthy series of procedural skirmishes, the plaintiff applied to the court for leave to amend its statement of claim by substituting an entirely new pleading. The defendant complained that its objections had still not been satisfied. The plaintiff's statement of claim was attacked on a number of grounds. Two of these are relevant for present purposes. The first was that specific parts of the claim were unintelligible and that the proposed amendment should be disallowed. The second was that the plaintiff had presented a global claim, which was of itself objectionable because it placed the defendant in a hopelessly embarrassing situation and in any event was bound to fail.

In relation to the plaintiff's claim for variations, after consideration of the (inevitably) massive and at times impenetrable detail submitted by the plaintiff the judge concluded that the defendant was entitled to complain about lack of clarity. He agreed that it was impossible to reconcile the various parts of the claim. Thus the pleading was embarrassing to the defendant. He ruled, however, that rather than striking out the pleading the appropriate course was to make the application for leave to amend conditional upon the provision of a schedule reconciling the conflicting elements.

As to the more general complaint, that the plaintiff had presented a global claim, which as such was inherently objectionable, the judge placed reliance on a previously unreported Australian case, *John Holland Construction & Engineering* v. *Kvaerner RJ Brown Pty.* Faced with a global claim similar to those discussed above, Mr Justice Byrne had pointed out the difficulties faced by a plaintiff in making good its allegations. He had however provided a warning to those seeking post *Wharf* to compel courts to strike out such cases.

'It is for the parties and not the court, even in a judge managed list, to determine how their case should be framed. It is not for the court to impose upon them a manner or form of pleading which it thinks better than their own.'

He continued

'The power of the Court to strike out is very limited. So far as here relevant it should only be exercised where the claim is so obviously untenable that it would be a waste of the resources of the court or the parties for the court to allow this only to be exercised after a trial.'

Judge Lloyd in *Rugby Landscapes* therefore formed a view based upon three principles

'(1) Whilst a party is entitled to present its case as it thinks fit and it is not to be directed as to the method by which it is to plead or prove its case whether on liability or quantum, a defendant on the other hand is entitled to know the case that it has to meet.

(2)   With this in mind a court may – indeed – must in order to ensure fairness and observance of the principles of natural justice – require a party to spell out with sufficient particularity its case, and where the case depends upon the causal effect of an interaction of events, to spell out the nexus in an intelligible form. A party will not be entitled to prove at trial a case which it is unable to plead having been given a reasonable opportunity to do so, since the other party would be faced at the trial with a case which it did not have a reasonable and sufficient opportunity to meet.

(3)   What is sufficient particularity is a matter of fact and degree in each case. A balance has to be struck between excessive particularity and basic information. The approach must also be cost effective. The information may already be in the possession of a party or readily available to it so it may not be necessary to go into great detail.'

Hence, while accepting that there was much in the defendant's complaints, he held that it would be wrong to refuse leave to amend; and, while it would be appropriate to attach conditions to some of the amendments with regard to particulars which should be given, he declined the defendant's application for leave to be refused and/or for the statement of claim to be struck out.

Accordingly, while the result may be said to differ from *Wharf* and to be closer to that in *British Airways Pensions Trustees*, it does seem that that the courts have settled that the manner in which a claim is presented shall only in the most extreme instances afford an excuse for striking out. What may be equally important from a tactical viewpoint is that, faced with clearly flawed or unintelligible pleadings, the Technology and Construction Court (as it has now been re-christened) will not shy away from telling a party in terms that a part or even the whole of its case appears to be badly flawed.

6.4   *Legal and practical consequences*

Of course, while the reported cases provide a fascinating insight into the way in which the courts have approached a series of poorly argued claims, none of them provide any kind of route map to construction professionals. It is impossible to say that any can be used in any given claim for delay, loss and expense to provide a universally applicable guide as to how a claim should be formulated or proved. This is not especially surprising – all were concerned with facts which were unique to the particular dispute. In this respect, however, the editor of *Construction Industry Law Letter*[6.10] makes a valid point: in the absence of any clear guidance, the defendant in any case will be tempted to argue that the case against him fails adequately to particularise the plaintiff's case.

From the perspective of a claimant, this raises an unpleasant prospect. Of course, if the comments made in Chapters 3, 4 and 5 have been heeded religiously, proving a claim should not present impossible difficulties. However, in any other instance, it is almost open season for a party to allege that the claim which he is compelled to meet offends against the strictures of *Wharf* and subsequent cases. It is no answer to say that the effect of the cases is to allow the claimant to argue that such claims should not be struck out, because to rely upon *Devonald Williams* and *GMTC* is to concede that a case has inherent weaknesses which, while it should be allowed to go to trial, may nonetheless mean it is met with short shrift when it gets there. In the circumstances, the only practical approach will be for a claimant to make the best possible attempt to comply with the broad guidelines from *Wharf* from the outset and attempt so far as possible to demonstrate the causal link between the events which led to the delays and the delays themselves.

This is plainly an unsatisfactory position for the industry, since it necessarily means that proving a claim with less than ideal material is turned into a marathon involving major battles by way of requests for particulars. However, in the absence of a genuine as opposed to imaginary retreat from the broad principles of *Wharf*, it represents the present state of the law, and the fact that the higher courts have thus far failed either to disapprove of *Wharf* or to offer some definitive guidelines on the degree of proof required in construction cases, the only answer appears to be to meet the case head on.

One of the most frequently voiced criticisms of the present uncertain state of the law is that it creates an atmosphere in which the construction team are necessarily more oriented towards dealing with disputes than towards confronting and solving problems.

It follows from earlier chapters that this perception is, in the writer's view at least, only partly justified. However, it is certainly the case that the avoidance of disputes in the post-*Wharf* climate has not been assisted by the differing provisions of the notice requirements of clauses 25.2 and 26.1 of JCT 80 which remain unchanged in JCT 98. It is noteworthy that the former requires the contractor to provide detail sufficient to allow the architect to ascertain precisely how and why delays have occurred (although this is more honoured in the breach than in the observance). The latter simply requires that the architect is provided with information to enable him to determine that relevant events have occurred causing loss and expense to be suffered: the architect therefore has the option whether to call for further information to enable him to make an ascertainment.

Some of the ramifications of this will be considered further in Chapter 8 but it is worth at this stage flagging two of the considerations which will confront parties:

- when should a respondent party claim that the other side should not be permitted to make a composite claim, and
- when should the architect call for further detail to enable himself or the quantity surveyor to ascertain loss and expense.

From a practical point of view this raises two important questions which are less matters for the theoretician than they appear at first sight. Firstly, whether it is self-defeating to say that the other party could apportion his claim among individual heads of loss – it follows from *Crosby* that if this point is carried to its logical limit it risks proving the other party's case for him. Secondly, this may be an important tactical consideration, particularly when it is plain that there are big differences between the values of different heads of claim. If one head of claim fails it may be important to carry out an apportionment in order to attempt drastically to reduce the scope of the remaining ones. The reverse side of these considerations, of course, is that there will be instances where it is very much in one party's interests not to mount a *Wharf*-style challenge to a claim. If it is plain that a claim in its present form has no prospect of success, it may be wise to resist the urge to force the other party's hand since the result of such an exercise may be to cause the claim to be recast in a more compelling form.

However, aside from these considerations there can be little doubt that the effect of *Wharf* has been to make the proof of delay claims more difficult in all but the best prepared cases.

CHAPTER SEVEN
ANALYSING THE CAUSES OF DELAY: PLANNING AND NETWORKS

7.1 Objectives

Chapter 5 deals with what to do and what not to do. Chapter 6 deals with the legal requirements of proving a claim. The purpose of this chapter and Chapter 8 is to put those lessons into a practical framework. In other words, how do we make the best possible use of what we actually have? This prompts two crucial questions – how do we plan the works in a way which will allow them to be executed within the constraints of available time and resources, and, in the event that delays occur, what exactly are we attempting to prove with this claim?

Few if any claimants will start with the intention of pursuing the matter to a contested hearing in which they will be required to prove that the events complained of led to the alleged delays.[7.1] Empirically, those claims where the claimant has set some clear goals are more likely to settle than those which are allowed to drift. Those goals tend to start with the planning of the works. Proper planning will facilitate the subsequent analysis of delays.

Those goals will be of three main types:

- Those arising out of the planning of the works. This subject occupies most of this chapter. Have the works been planned in a way which will permit them to be executed within the available time and resources? It is a truism to say that, if the works have not been properly planned, it will come as no surprise if those works then suffer delays.
- Those concerned with the degree of proof which can be achieved. How far can we go to answer the checklist in section 5.2? What resources can we put towards achieving this and what costs can we sensibly devote to this?

- Those concerned with the end result. Given the quality of the claim which can be produced, what can realistically be achieved? Just as disputes are frequently the product of a failure to appreciate the true scope of the parties' obligations, so litigation or arbitration frequently result from a failure by one or both parties properly to evaluate the real strength or weakness of a claim.

This is not a complex exercise. Often, however, it involves some unpalatable truths being recognised. The questions posed in the previous paragraph will frequently lead the party concerned to conclude that his planning of the works, his performance of the works, his administration of the contract and the records he can marshal in support of his contentions will simply not be sufficient to enable him to produce anything remotely persuasive and, for the reasons touched upon above, any attempt to do so may not only run the risk of long drawn out proceedings but it may hamper his prospects in relation to other aspects of the contract.

The lesson from Chapter 6 is that many claims will be criticised and may fail, not because of a lack of information but because that information is not assembled and put forward in a way which enables the receiving party or ultimately a judge to discern what the claimant is attempting to prove.

7.2   Strategic planning

The sample scenario

The sort of issues facing many potential claimants are best illustrated by an example. This has been chosen on purpose as a factual scenario in which the claimant has bad points as well as good and where there are also 'political' considerations to account for. The issue of planning the works has been deliberately simplified.

- The claimant is a main contractor undertaking a social housing project on behalf of a housing association which is being partially funded by the local authority as part of the replacement of the housing stock. The works are being carried out in accordance with the terms of JCT 98 with contractor's design (WCD 98). The two largest elements of the works are brickwork and the groundworks which include the roads and landscaping. The programme for the works comprises a simple bar chart in which the majority of items are shown sequentially – the start of one occurring when the previous activity ends.

- During the early months of the project, a number of problems arise:
  - o Firstly, an unseasonable period of wet weather causes the site to be flooded which delays the groundworks.
  - o This coincides with the employer deciding to change the configuration of one of the blocks, replacing flats with maisonettes and thereby altering the configuration of the drainage runs and the access to the entrances.
  - o He also lets it be known that the contractor's planned method of gaining access to the site will not be possible because the local authority is refusing to relax a prohibition against construction plant using a particular road leading to the site.
  - o Following a series of clashes with the employer's agent, the contractor's site agent leaves.
  - o There is a gap in the records while a replacement is found. A delay of eight weeks occurs.
- The brickwork sub-contractor is not informed of the extent of these delays and had been told to order materials which are to come from a particular local brickworks. The bricks are delivered to site and application is made by the sub-contractor for payment under the provisions of his sub-contract. Disputes arise as to the amount to be paid after some of the stock becomes saturated. This problem is exaggerated by a series of early frosts. This delays both the drainage and the brickworks. Problems with the latter are increased because of cracking to some of the wet bricks which have been inadequately protected, and the inability of the brickworks to supply replacements to order.
- To minimise the effect of these delays, the employer's agent instructs the contractor that, rather than waiting for the entirety of the drainage to be completed, he should re-sequence the brickworks to follow immediately upon completion of each of the drainage runs. This decision misfires because it diminishes the productivity of the brickwork sub-contractor. Combined with the fact that the works are progressively being pushed into the winter period, the rate of progress slows further.

Fig. 7.1 shows a comparison between the original programme for the groundworks and brickwork and the rate of progress actually achieved together with the main events identified above.

This scenario illustrates the following:

- Some of the delays are the fault of the contractor, others are the fault of the employer; some carry an entitlement to direct loss and expense, others do not.
- Some of the delays run in parallel ('concurrently') with other delays while others do not. Some depend in whole or part upon previous events while others are independent.
- Unlike certain manufacturing processes, where the start of one operation is dependent upon the completion of the previous task, construction generally involves inter-related operations and so delay to one task will not necessarily delay following operations.
- The problems are exacerbated by the way in which the works have been planned which leaves little or no scope for most of the activities to slip. To cope with the problems a large scale re-sequencing of the works is necessary.

The temptation will be to argue that because the total delay amounts to a particular number of weeks – the amount of delay shown by comparing planned with actual finish, this therefore marks the extension of time to which the contractor is entitled. This is unrealistic because it will be said, with justification, that it is impossible for the contractor to say that the whole of the period of delay can be attributed to matters entitling him to an extension of time. Furthermore, the way in which the works were programmed means that simply adding the separate effects of the various causes of delay does not reflect what actually occurred, nor what had actually been planned. This demonstrates the point which was made in cases like *Wharf Properties* v. *Eric Cumine Associates* (1991) 52 BLR 1 – simply aggregating the delaying events together, and sifting out those which are inconvenient to the contractor does not provide a basis for determining what period of delay has been caused to the contractor. To use the expression favoured by the court in *Wharf Properties*, it does not provide an 'agenda for trial'.

Hence the importance of determining the relative importance of different delaying events and placing these events in the context of what had originally been envisaged and what occurred in fact. An

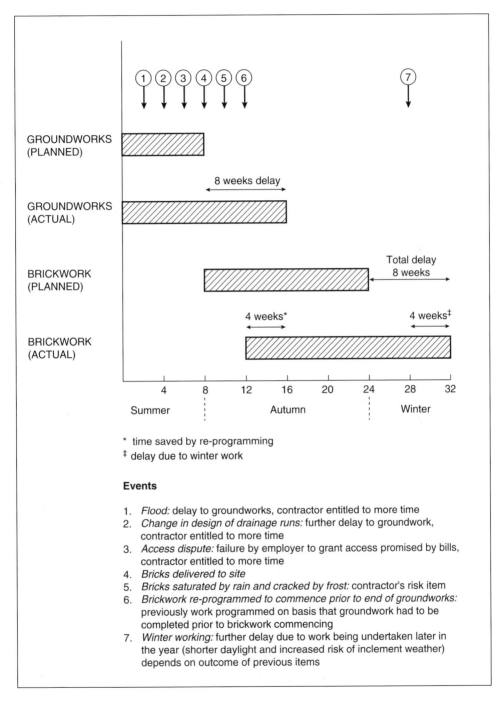

Fig. 7.1 Programme of works: bar chart.

example is provided by the firebells referred to in *ICI* v. *Bovis Construction Ltd* (1992) 32 Con LR 90 considered in Chapter 6. The result of the aggregation and sifting process referred to above was that the court was there being asked to accept that the moving of a firebell would have the effect of delaying an entire project. In the absence of evidence to support this proposition, the court had little difficulty in rejecting the notion. What it also shows – and if anything this is more important – is that any form of delay analysis is only as good as the data on which it is based. The question is always – 'delay to what?'

The programme as a planning tool – some terms

Going back to the programme at Fig. 7.1, the original programme for the works can be shown with the simple inter-relationship between activities plotted on the original programme. In Fig. 7.1 the original programme showed the brickwork commencing only after the groundworks were completed. Following events 1–3 the works were re-programmed to provide for the brickwork commencing before completion of the groundworks, with the intention of making up four weeks of delay. Thus the total period of delay amounts to eight weeks, compared against the cumulative total of delays of 12 weeks. Such a methodology can show not only the order in which the main contractor intended to carry out the works but also the milestones which had to be reached in one activity before the contractor could move on to the succeeding activity. It can then be seen how the delaying events affected that sequence. This allows the determination of which events can be said to be the responsibility of the contractor, and which are matters for which he can properly claim an entitlement to an extension of time, and most importantly, which of the delaying events run concurrently with others, and which can be said to be critical to the progress of the works as a whole.

This introduces two crucial expressions, '*critical* activities or delays' and '*concurrent* activities or delays'. The former are matters where delays will in themselves cause the progress of the works to be delayed. The latter run in parallel with other programme activities and delay to the progress of these activities will not of itself cause the progress of the works to be delayed. These are expressions which will be considered in greater detail below.

In the simple example in Fig. 7.1, the programme for the works was produced on the straightforward basis of considering what had to be done in order to allow the works to move on to the next stage. The task of determining the critical and concurrent delays is one which in our simple example can be undertaken subsequently and without great difficulty. Plotting the succeeding critical activities will allow the parties to determine the 'critical path' for the works. A third party looking at the delays which have occurred will be able to see exactly where the problems have arisen. He will also be able to determine the points which he needs to address in deciding where responsibility for those delays lies. He has an agenda for trial.

Regrettably, and as the decided cases show, the history of delay claims is less simple. In large part this is because construction works are seldom as simple as the example given above. There the task of determining which activities are critical and which are concurrent is straightforward and obvious. In most projects, however, the criticality of events will be less obvious, and, indeed, while there will be identifiable critical events, certain activities will only *become* critical in the event of certain conditions not being fulfilled.

This complexity is also attributable to the methods adopted in attempting to prove delays, in which simplicity has often been abandoned in favour of needless over-elaboration – often for the reason that the methods do not exist which will enable the claim to be proved by simple means. In reality this is also sometimes a way of attempting to disguise the fact that the simple approach may produce an unpalatable answer.

Drawbacks with the use of bar charts

While the sort of simple bar chart shown in Fig. 7.1 will be effective in the simplest instances, its value diminishes in more complex projects particularly those featuring complicated inter-relationships between activities. The first drawback can be illustrated by Fig. 7.2.

- Activity 1 must be completed before Activity 3 can be commenced.
- Activity 2 must be completed before Activity 6 can be commenced.
- When Activity 2 is 75% complete, Activity 5 can be commenced – thus limited overlap is possible between Activities 2 and 5.

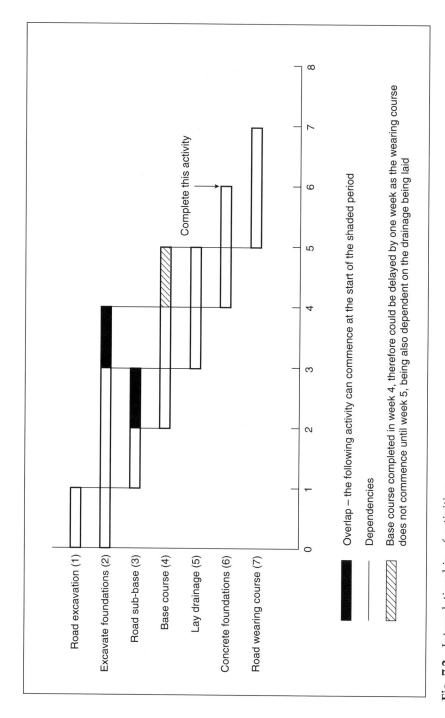

Fig. 7.2 Interrelationship of activities.

- When Activity 3 is 50% complete, Activity 4 can be commenced.
- Although Activity 4 completes by week 4, Activity 7 which depends on it does not commence until week 5 because it also depends on Activity 5. Activity 4 can therefore be delayed by a week without delaying following Activities.

In simple construction works, these relationships can be shown by the sort of linkages shown in the diagram. However, if the activities shown in Fig. 7.2 simply comprise one block out of several, it will become more difficult to show how these matters inter-relate.

The second drawback is a function of the planning process. The sequencing of works depends upon balancing three considerations:

- Logic – decisions on the way in which the works are to be carried out. In the example above most of these decisions such as the fact that the foundations and drainage must be completed before the foundations can be concreted, take themselves, but in more complex works, this will not always be the case.
- Time – the period which a particular activity will occupy.
- Resources – the assumption that the means will exist to carry out a particular activity and that particular levels of resource will be devoted to it to enable the work to be carried out within a particular time span.

The planner, called upon to produce a programme for a project cannot decide on these three matters simultaneously but necessarily has to take these decisions sequentially. Indeed, in the majority of works, a decision will need to be taken to prioritise the three considerations.

Once again the relationship between time and money appears – in planning works in which the dominant consideration is the scarcity of particular resources, it may be necessary to extend time periods for particular activities to enable the resource to be deployed effectively, or alter the logic of the job so that activities are executed consecutively rather than concurrently. If bricks can only be supplied at a certain rate the time for completing all the bricklaying has to be extended and such work programmed sequentially, house by house, rather than all being done at the same time. Taking the project as a whole a fixed date for completion may mean the carpenters/joiners following the bricklayers very closely rather than having that task delayed until the other is complete.

Where, by contrast, the dominant concern is time, it may be necessary to use more of particular resources than otherwise to enable particular activities to be compressed.[7.2] Again, the logic of the works will be affected. In both cases the inter-relationship of these considerations will affect the cost to the consumer of the end product.

Strategic and tactical planning

As we have seen from Chapter 2, and particularly the conclusion to Section 2.3, the success of a project may depend in large part upon events which occur before the parties get anywhere near site. Lockyer and Gordon[7.3] adopt a military analogy by distinguishing between 'strategy' (that which serves the needs of generalship) and 'tactics' (plans which are made when in contact with the enemy). They make the valid point that the best strategic plans are those which are made before the campaign begins and which allow the matter to be carried through to a successful conclusion, but recognise that inevitably circumstances will arise requiring tactical decisions which will be successful if and only if they are made within the context of the overall strategic plan.

Thus the preparation of the programme will require the planner to balance logic, time and resource. The starting point will generally be the identification of the order in which the works are to be undertaken, the logic, periods of time are then allocated to these activities, although these periods (together with the precise inter-action between the activities) will be dictated by the available or necessary resource required to execute them. This process of adjusting the logic, time and resource required will continue until the scales balance and the programme shows completion being achieved within the allowed time using an available or affordable level of resource. This will show the total project time (TPT).

This is the method by which estimators have determined whether works can actually be accomplished within the required time and budget. Significantly it will also be invaluable as a tool for determining whether works which have suffered delay, ever could have been executed within the planned logic, time period and resource availability.

While in many instances the production of the programme will depend upon the use of one of the various software packages which exist, it is a fatal error to believe that the computer programme will dictate the programme for the works. On the contrary, the software (however sophisticated) will do no more than assist in the presentation and ordering of activities within the constraints of logic, time and resource which are available to the person using the programme.

Accordingly it is necessary to look in a little more detail at the methodology of programme preparation and then at the way in which the same techniques can be adapted to analyse the causes and effects of delays .

7.3 Project network techniques in programme preparation

The network

While a simple bar chart will be a useful tool, in many projects its value will be greatly enhanced if it is produced in conjunction with or as a product of a more sophisticated network analysis of the component parts of the works. The value of the network analysis if properly produced is to enable the parties to plan with precision how the works are to be carried out, and self-evidently to analyse retrospectively how delaying events have actually impacted upon the progress of the works.

The most commonly used method of network analysis is known as the 'activity on arrow system', commonly referred to as 'critical path analysis' (CPA)[7.4] or 'programme evaluation and review technique (PERT).[7.5] This approach has two ingredients:

- Activities – a part of the work, occupying a time period, which must be carried out.
- Events – almost invariably the start or finish of an activity or group of activities, also referred to as milestones or nodes.

Activities are represented by arrows and events by some convenient shape – circles or diamonds are the most commonly encountered. As with bar charts, time flows from left to right and successive events and activities have ascending serial numbers. Where a par-

145

ticular time span is required for an activity this can be shown on the arrow itself which will often save time otherwise needed to check the periods shown on the horizontal axis. If for some reason not forming part of the network itself, a particular activity needs to be completed by a particular date, this can often be shown most conveniently by an inverted arrow at the end of the activity arrow. An example of these matters can be seen in Fig. 7.3(1) and is developed in Fig. 7.3(2).

In fact, most projects are planned on the basis that the time

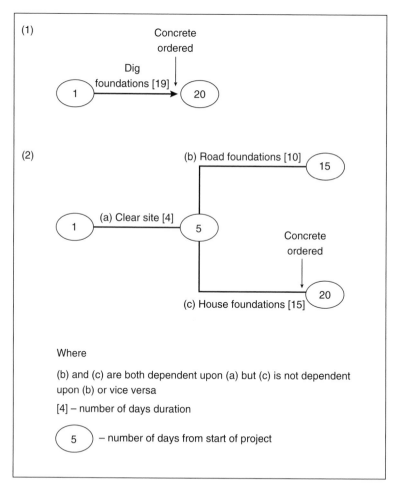

Fig. 7.3 CPA activity on arrow system: a simple model.

allocated to the particular activity is greater than the time which is likely to be needed to complete the activity. This is for the good reason that some 'slack' or 'float' should prudently be provided to allow for the sort of unpredictable events which occur on any project. It is also frequently the case that, however skilled the estimator, he will often be able to do no more than make an inspired and educated guess as to the length of time likely to be needed to complete a particular activity. It goes without saying that the usefulness of any network programme will depend upon the realism of the activity durations. Equally, the first line of attack on any network programme is the contention that the programme could not have been operated in practice and that the delays occurred not because of an act of prevention but because the estimated activity durations were unrealistic.

7.4   The critical path

The total project time (TPT) is the shortest time in which the project can be completed. To establish this, it is necessary to identify earliest starting and finishing times (EST and EFT) for each activity and then to identify latest starting and finishing times (LST and LFT) for each activity. Two further concepts need to be identified, earliest event and latest event times (EET and LET). These are respectively the earliest date when the event can be realised and the latest time by which the event *must* be completed in order to achieve the total project time.

This is shown in practice in Fig. 7.4. Activities A, B and C represent the excavation of three blocks (A, B and C) forming part of the same project. By convention the event circle is divided such that the left hand semi-circle shows the event number, the upper right hand quadrant shows the earliest event time (EET) and the lower right hand quadrant shows the latest event time (LET). Thus:

- Activity A (event node 1) has a duration of 5 weeks and an EET of 1 week.
- Activity B (event node 2) has a duration of 7 weeks and an EET of 2 weeks.
- Activity C (event node 3) has a duration of 10 weeks and an EET of 3 weeks.

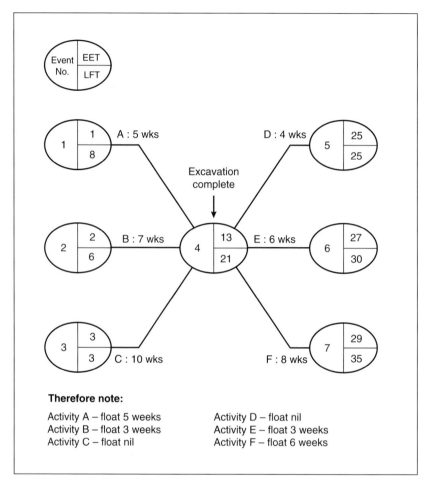

Fig. 7.4 CPA: a more complex model

Hence

- Activity A has an EST of 1 week and an EFT of 1 plus 5 weeks i.e. 6 weeks
- Activity B has an EST of 2 weeks and an EFT of 9 weeks.
- Activity C has an EST of 3 weeks and an EFT of 13 weeks.

The earliest date at which all the excavation activities are complete is therefore week 13. Thus the LET for event 3 is 13 weeks and the EET for event 4 (perhaps the time the concrete is ordered for as in

Fig. 7.3(1)) is also week 13, being the latest of the EFTs of the activities leading into it.

Activities D, E and F are the concreting of foundations for the three blocks. If therefore

- Activity D has a duration of 4 weeks and an LET of 25 weeks
- Activity E has a duration of 6 weeks and an LET of 30 weeks
- Activity F has a duration of 8 weeks and an LET of 35 weeks

a retrospective survey will establish latest start and finish times (LST and LFT) so that

- Activity D has an LFT of 25 weeks (i.e. the LET of the event node the activity arrow leads into) and an LST of 25 minus 4 (i.e. 21 weeks)
- Activity E has an LFT of 30 weeks and an LST of 24 weeks (30–6)
- Activity F has an LFT of 35 weeks and an LST of 27 weeks (35–8).

The latest date for event 4 to be realised is the earliest of the LSTs for the emerging activities, namely week 21.

What we see is that if we consider Activities E and F, the latest that either can start is week 21 and the latest that Activity E can finish is week 30 and the latest that Activity F can finish is week 35. However, Activity E has a duration of 6 weeks. There is therefore a *float* period of 3 weeks. This additional period can be used up without increasing total project time. Similarly, Activity F has a float of 6 weeks which can be used without increasing total project time. Provided that the activity is complete by the latest finish time (LFT) it does not matter whether the works exceed the planned duration.

Contrast Activity D: there is no spare time and therefore if this activity starts late or exceeds the planned duration the total project time will be exceeded. Activity D may therefore be said to be *critical* and the *critical path* lies along Activity D. This demonstrates a fundamental rule: the critical path lies along the activity with nil float or the activity with the least available float.

It will also be appreciated that the float possessed by one activity may impact upon a succeeding activity. This is shown in Fig. 7.5.

The earliest that Activity A can start is week 9 and the latest it can finish is week 32. The *available time* is therefore 23 weeks. Since the duration of this activity is only 15 weeks, the float for this activity is

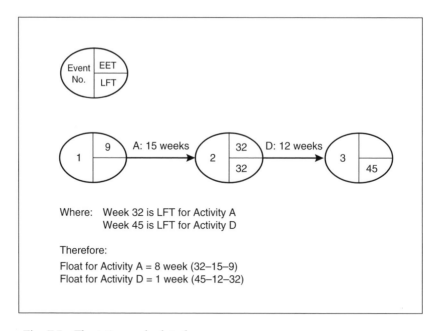

Fig. 7.5   Float time calculated.

8 weeks. This period can be used up without delaying later activities or extending the total project time, but if this period is exceeded the total project time will be extended and the critical path for the project will change such that the critical path runs through this activity. How will this impact on Activity D? The maximum time which will be available for this activity is 13 weeks. The duration for this activity is 12 weeks and therefore the float is apparently one week. But, if the whole of the available float for Activity A and more is absorbed, the effect might be that Activity D will not be able to start until week 33. Although Activity D can still be undertaken within the planned project duration, the float will have been absorbed by the preceding activity.

This requires us to consider the further concept of 'free' float – that by which this particular activity can be extended without affecting the float of other activities. Thus in Fig 7.5 there will be a *total* float of 9 weeks but a *free* float of 7 weeks – above 7 weeks and this activity will start to eat into the planned float for succeeding activities, i.e. beyond event node 3.

7.5   *Resource analysis*

In the majority of construction projects, time is regarded by the client as inflexible. The client will wish to see the works completed by a particular date, perhaps because this is the date when he hopes to allow a tenant into occupation, or when he is committed to allow a particular public works project to be opened to the public.[7.6] The competitive tendering process means that in many cases the successful party is the one who makes the most efficient analysis of the resources which will be needed to complete the works within the available time.

The tendering process is intended to identify errors in tenders where serious under-allocation of resources has caused the bid price to be mistakenly low and there is doubt whether the works actually can be executed within the resource budget.

Resource allocation

Precisely calculated, the available resource will exactly equal the required work such that the resources are working at maximum safe operating capacity. This is rarely desirable – a small miscalculation will lead to the resources allocated being overloaded such that it will be impossible to carry out the particular activity within available time so that delays become inevitable. Just as engineers when calculating foundation loading would be wise to build a factor of safety into their calculations, so the prudent planner should build a degree of spare capacity into his estimates. Often this is in order to make due allowance for resources which cannot be estimated with complete precision.

A simple example explains this: during the course of the refurbishment of part of an underground railway, the contractor estimated that his workforce would be able to splice a particular number of cables in a given period using a particular number of men and a specialist splicing tool. He priced the works on the basis of this projected output but reckoned without the fact that the works were to be carried out in the platform inverts in poor light, with high ambient temperatures and high levels of dust and debris accumulated over a century. This impeded the operation of the equipment which had been designed for use in clean cool conditions, and hampered his workforce whose work was less efficient. Unsurprisingly, the activity was delayed.

This example also shows the danger inherent in assuming constant rates of output or progress and assuming that these can be maintained throughout the activity. Particularly where software based estimating or programming packages are used care should be taken to allow sufficient safety margins to avoid the sort of difficulty envisaged above. Similar care should be employed to avoid the mistaken assumption that what two men can do in twelve hours, four men will necessarily accomplish in six.

The process will also require consideration of whether the whole activity can be executed more quickly by the use of more resources. This is not only a question of whether more resource equals more progress; but also whether the tasks which make up the activity can be executed all at once. This is the same issue as the programming of the works as a whole. Just as it is impossible for each of the activities comprising a project to be executed at once, so it will often be impossible to carry out the whole of an activity at once.

Applying this to a network analysis or to a bar chart shows how the interplay between time and resource can be used to best advantage.[7.7] In projects where the primary consideration is time, the estimator can identify the units of resource which will be needed to complete each activity. This can be plotted on a simple labour or plant histogram to show what resources will need to be deployed for each activity. Fig 7.6 shows a simple example and it will be seen that the available resource is shown as a dotted line. Should this reveal (as here) that the required resource to undertake the works cannot be deployed – either because it does not exist or cannot be obtained or because that quantity of a particular resource cannot be deployed to produce that level of progress – the required time for an activity will need to be adjusted. Where the primary consideration is the availability of resources of a particular type, the converse question arises – can the activity be accomplished within a desired time using the available resources.

This process will also show that in many cases the ideal use of resources may require a degree of flexibility which is not achievable. If it appears that the contractor will need to vary his workforce from 100 in week 1 to 5 in week 2 and 250 in week 3 he will probably be obliged to accept that this level of limitless flexibility is unlikely to be practicable.

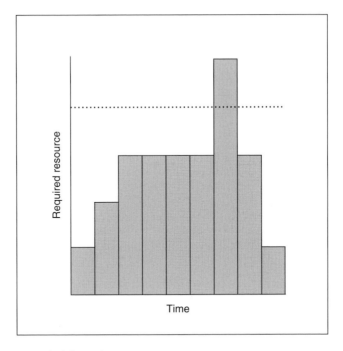

Fig. 7.6 Simple labour histogram.

We can show this by representing the required resource loadings as a histogram. An example is shown in Fig. 7.7. As a rule of thumb, such a histogram should avoid pronounced peaks and troughs. While it may be that from a theoretical perspective the sequencing of the works will call for the use of the maximum possible level of a particular resource in one week followed by none in the following week, it will rarely be the case that this actually represents an efficient way to use resources. It is unlikely that the contractor will actually be able to juggle his resources between contracts in a way that enables him to vary the levels of resources between jobs and maintain maximum theoretical efficiency.

For this reason, if the histogram derived from the network shows that the works are planned in a way which produces undesirable peaks and troughs it becomes necessary to look again at the network. The results and actions taken will naturally depend upon the particular circumstances. However, this does indicate an important principle – namely that planning tools are not ends in themselves

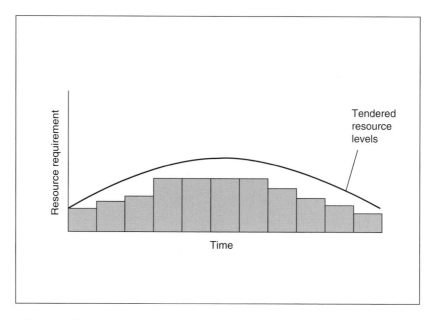

Fig. 7.7 Required resource loadings histogram.

and that if the results achieved first time around are not appropriate, the tools should be reconfigured in order to produce a result which is better suited to the proper execution of the works. In particular care should be taken to see whether there are any periods of float time which can be utilised in order to reduce particular resource peaks or troughs. Obviously the execution of a particular activity with a lower level of resource over a longer period will not increase total project time (TPT) if the increased time represents float.

While the network and the histogram derived from it will be invaluable in planning the work, and indeed in analysing the execution of the work after it has been carried out, care is needed in any attempt to re-programme the works to ensure that the production of revised programmes does not simply become a historical exercise to plot the delays which have occurred since the previous version; and further to endeavour to create a revised programme which will actually be a valuable aid to completing the works.

7.6   *Overview*

The network and the bar chart used together

In Section 7.2 we looked at the difficulty in attempting to *plan* work simply by plotting the works on a bar chart. Section 7.4 shows the use of the network as a much more flexible tool in planning the works. However, the network is a less useful device for demonstrating at a glance how the works are to be executed. The bar chart has the great advantage of *immediacy*. In a site context, the overwhelming majority of the site staff will be able to appreciate what follows what.

Furthermore, there is no difficulty in translating a network into a bar chart. Calculation of duration times from the activities and plotting these on the time scale of a bar chart is extremely easy. However, while the textbooks on the use and production of networks suggest that it will then be possible to show the interdependency between activities making up a network on a bar chart, this approach tends to produce unnecessary confusion. The bar chart and the network serve different purposes, the former shows the order in which the works are to be executed and the latter demonstrates the interdependency between the activities comprising the works. Put another way, in the vast majority of construction projects, the bar chart will be the tool used by those *executing* the works, the network will be of most use to those *controlling* the works.

Chips with everything

This chapter has shied away from consideration of the use of computerised programming techniques. This has been for the simple reason that the various available packages are still servants rather than masters. They are only as good (or as bad) at programming the works as the person inputting the data and assimilating the results. Even the very best software packages are still susceptible to producing silly results if provided with the wrong information.

Hence it is essential to understand the basics of project network techniques even where the nuts and bolts of the programme are to be produced by computer. Not least this will enable the planner to see what his software programme has produced and analyse it for obvious illogicality, which in turn will act as a check on the integrity and logic of what has been input.

As importantly, and this is an issue which will be examined in more detail in the following chapter, it will allow the planner to respond appropriately to events which cause delay to the works. This will allow the consequences of the delaying event to be factored into the re-programming of the works in such a way as to produce a revised programme that provides the contractor with a feasible and practicable route through the remainder of the works. The alternative, which has unfortunately been seen on a number of civil engineering projects in the past decade, bears an uncomfortable resemblance to the legend of the sorcerer's apprentice. The result is, at best, a programme which provides a flawed view of how the works were planned and is of little real help in assisting the parties to cope with delays.[7,8]

Conclusion

This chapter has looked at the techniques to be employed in planning a project. The next chapter deals with analysing the causes of delay. The techniques and terminology employed in planning are equally applicable to the consideration of delays. There is a temptation, though, to plunge into the causes of delay without paying due regard to how the works were originally planned. There is very little merit in determining that the works were delayed by a series of matters if consideration of the originally planned sequence for the works shows that these delays were inevitable given the way in which this exercise was done.

Consequently it is hardly surprising that many claims are met with the argument that, while the works to which they relate may well have been delayed, they could not in any event have been constructed during the programmed period given the resources allocated. Hence, even where the planning for the project has been carried out on a much more simplistic basis than that described above, the first task of the person charged with preparing a claim is to 'prove the tender' – demonstrating that it was possible to construct the works in the time permitted and using the resources allocated. This comprises a retrospective exercise to check on the planning of the project using the project network techniques described above.

In the context of planning the project, the usefulness of project network techniques is that they allow the parties to predict the effect on concurrent and consecutive activities of particular aspects of delay. In the context of analysing the delays which have actually occurred they permit the claimant to map out a logic within which the works were to be carried out. The application of the delaying events to this logic *should* provide a compelling explanation of what actually happened and why.

Examination of the decided cases suggests that this approach is seldom employed in those cases where the courts have criticised the presentation and methodology of particular claims. The curious feature of this is that to ignore project network techniques as a means of proving a claim seems inevitably to create a great deal more work and to produce a markedly inferior result.

CHAPTER EIGHT
DELAY ANALYSIS

8.1   Introduction

In the simplest contacts the question 'what happened?' will be answered by an account by someone with first hand knowledge of the facts – usually the builder, who will be able to give an account of the events which caused the works to take longer than had been intended. In slightly more complex jobs, where the works have been planned using a simple bar chart, it will be possible (again using first-hand evidence) to explain what had caused the individual bar lines to be extended and perhaps why the prolongation of one bar line has affected another.

As the works become more complex, two problems start to emerge. The first is to link the accounts of delaying events to the actual delays. The second is that this process increasingly involves a degree of subjective analysis. All too frequently, the notion of linking the events to the results becomes lost. The claim becomes an amalgam of a series of facts, some agreed, others contested, and an assertion that these matters caused delays. The causal link is lost or, to be exact, is never made. Examples of this are provided by some of the cases and particularly *Wharf Properties* v. *Eric Cumine Associates* (1991) 52 BLR 1 and *ICI* v. *Bovis Construction Ltd* (1992) 32 Con LR 90. In both it is easy to see that the claiming party failed utterly to demonstrate that the large volume of factual material actually explained why the works had been delayed.

Accordingly it is necessary to consider the techniques which may be employed to link events to their consequences. The starting point for this is the project network techniques (PNTs) which are the subject of the previous chapter. Delay analysis is really no more than the retrospective application of the planning techniques which are used to plan the job in the first place. As such this need not be a complex process. While delay analysis is rapidly assuming the status of a science in its own right in the United States,[8.1] in this country it is a process which is understood by too few of those construction professionals whose livelihood is earned by the production of claims. Too often the process of analysing (or even merely accounting for) delays is approached in a way which accentuates the complex (with the result that) the facts are rendered utterly unfathomable) in the hope that the tribunal will conclude that, simply because they cannot understand it, there must be some merit in what is being advanced.

This chapter therefore concentrates on simplicity of approach. It is acknowledged that a great deal of what follows assumes that the writer of the claim will have access to records that are sufficiently complete to enable the substantiation of such facts as are necessary to prove the underlying contentions. The importance of adequate record keeping is dealt with elsewhere but at risk of repeating the adages of Chapters 2 and 3, it is worth stressing that delay claims are primarily an exercise in showing why particular things took longer than they ought to have done. In almost all cases it will be impossible to do this unless the data exists with which to show when the crucial events actually took place.

This can be seen from consideration of the component steps in the analysis.

(1) *Tender analysis* – could the works actually be constructed using the resources allocated to the task by the contractor in his tender?

(2) *Programme analysis* – using the resources allocated in the tender was it possible for the claimant to construct the works in the manner envisaged by the programme and was this actually the way in which the contractor intended to build the works?

(3) *Event analysis* – identification of the events which caused the works to be delayed and analysis of their effect both on each other and on the completion of particular activities or the works as a whole.

Of these steps, only the final part of step 3 can be said to be entirely evaluative, that is to say dependent upon some form of subjective input to determine how the various matters under consideration affected the progress of the works. Steps 1 and 2 are largely (and in most cases entirely) capable of being proved by evidence of fact to show the integrity of the tender, the feasibility of the programme and the actual intentions of the contractor.[8.2] The identification of the facts giving rise to delays should also be an entirely factual exercise. It acquires a subjective quality and depends upon opinions or speculation only when records fail to demonstrate how particular delays occurred. As we shall see, the greater the dependence upon opinion or indirect evidence to establish the facts, the harder it will be to prove that these matters actually caused the delay which is complained of.

In many cases, particularly those where activities follow in a largely sequential order, the establishment of the facts will substantially accomplish the task of proving the causes of delay. This is shown in Fig. 8.1. which shows the construction of a prefabricated bungalow by a small builder.

- Activity 1 is the digging of foundations. Until this is completed the floor slab, Activity 2, cannot be poured.
- Activity 3 is the erection of the carcass of the prefab. This must be completed before Activity 4, the installation of glazed units, can be commenced.
- In turn, these must be finished and the structure made weathertight before the electrical conduits, Activity 5, can be commenced.
- While in theory the erection of partitions and the execution of finishings could be commenced before the completion of the electrical works, the availability of limited resources prevents these tasks from being carried out other than in sequence.

The glazed units do not arrive when planned. The fact that they do not can be easily established, and the delay can be attributed solely to this cause and this is shown in the comparison between the 'as planned' and as built bar charts. The only area which may give rise to argument is whether this delay occurred as a result of fault on the part of the builder or his client.

8.2 Tender and programme analysis

The starting point in any exercise of this type is to show that the claimant could actually have built the works in the intended manner. This depends upon a legal and a practical question.

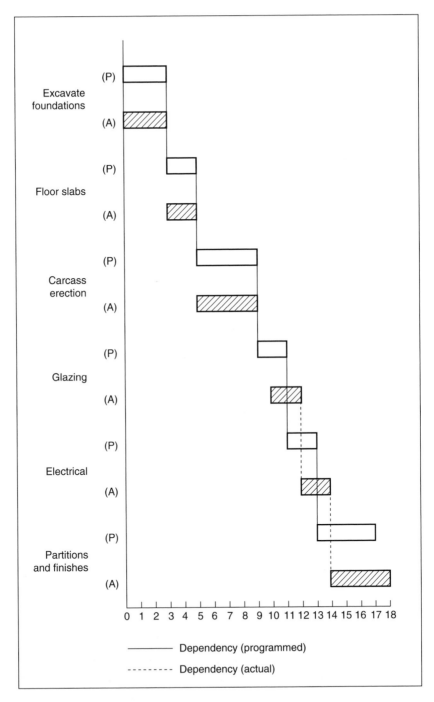

Fig. 8.1 Prefabricated bungalow: building programme.

The legal question can be dealt with simply. Is there any reason why the contract restricts the claimant's ability to build the works in his chosen manner? If the contractor planned the works, for example, on the premise that access to particular work faces would be exclusive and uninterrupted, notwithstanding statements in the contract specifically limiting his entitlement to sole or continuous access, it seems likely that the claim would get off on the wrong foot.

The practical question is whether the works actually could be built within the time and programme restraints given the resource allocation made. An interesting feature of the American decisions is the quality of the analysis carried out on whether the claimant actually could do that which he had contracted to do, or whether in fact the underlying cause of both the delays and the requirement for additional labour or materials could be put down to the fact that the job was underbid in the first place.

This should be an entirely factual question. If the contractor's tender calculations show that in assembling his bid he allocated particular levels of labour and resources to the job and that this was realistic, the question will be answered in his favour. This is particularly the case in contracts tendered on the basis of bills of quantities where the work and material content of particular activities should be capable of ascertainment with a reasonably high degree of precision. Although this is frequently an issue which will be addressed by the parties' experts, the weight to be attached to the opinion of the respective experts will generally depend upon whether their opinions derive from verifiable facts, rather than mere hypothesis.

Thus even in the simplest contacts where the works may be described in quite vague terms provided that there is certainty as to what is to be constructed, the question is whether the contractor can demonstrate that the planned resources were adequate to enable him to construct the works in the permitted period. Even in design and build contracts, where the Contractor's Proposals may be little more than a statement that the contractor will build what is described in the Employer's Requirements, the issue is whether behind this he has the ability to demonstrate the method and resource needed to fulfil the contract.

In more complex contracts the use of project network techniques as a planning tool will be invaluable in showing that the works were planned with sufficient detail to enable the contractor to demon-

162

strate that they could have been built for the price agreed. Where the works were not actually planned by this method it may still be possible, given sufficient estimating records, to recreate the planning process applying project network techniques to demonstrate the adequacy of the tender. Figure 8.2 shows two examples of this. Both are limited to a single activity. In the former it is clear that the labour required will be sufficient to allow the works to be carried out within the allowed period. In the latter they will not. It will be noted that the approach is complementary to that used in Chapter 7.

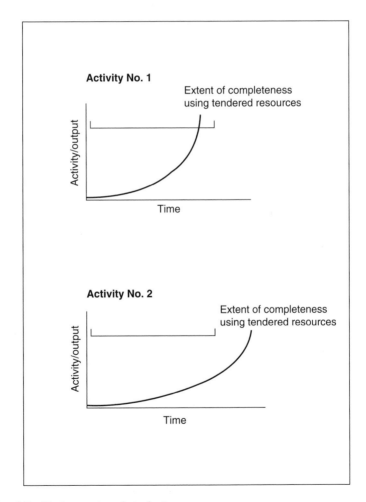

Fig. 8.2 Project network techniques.

In preparing his tender the skilled estimator should be able to take account of:

- The extent to which the prevailing working conditions will be more or less advantageous and hence whether productivity will be more or less than that to be expected in average conditions.
- Whether the sequencing and progress of the works will be affected by matters such as reduced daylight in winter, or the likelihood of particularly adverse weather conditions at exposed sites.
- The extent to which certain activities will be affected by the learning process which is inevitable in certain activities under-taken in an unusual way or in unusual conditions.
- The impact of other trades – to what extent is this activity reliant upon exclusive access to work faces.
- The relationship between time and resources – as we have commented in Chapter 7, an increase in (for example) labour will seldom produce a proportionate reduction in time required to accomplish that activity.
- The opportunity for overtime.
- The effect of altering levels of supervision.

While the estimator has to use his judgement to assess the likely effect on his productivity of these matters, the analyst attempting to test the integrity of the tender has the advantage of being able to see whether the assumptions on which the job was tendered proved justified.

Assuming therefore that the estimating records are available and are reasonably detailed, the only real debate may come where it is argued that in fact the requirements of the works were such that the works as tendered could not be completed within the contract period. Where the issue is quite finely balanced, this point produces a dilemma which explains why it is an argument generally only encountered in obvious cases. The dilemma is that if the issue is indeed finely balanced, there is a clear risk that the tribunal will conclude that the effect on the claimant's progress was marginal *unless* faced with clear evidence that the shortfall actually rendered the works impossible. It is worth noting that this is not the sort of issue where complex expert evidence demonstrating the theoretical impossibility of a particular activity is likely to be rewarding. Robert

Fenwick Elliott notes that what is required is fact, a statement of what actually happened.[8.3]

The obvious basis on which to attack the adequacy of a tender is to show that the resource levels allocated to particular activities are inadequate – hence, the works could not be built for the price. This may not be simply because the work content of particular activities exceeded the quantity allowed in the tender. It is common to find that the works have been tendered on the basis that particular output levels would be achievable but were not because the opposing party sought to place restraints on the claimant's abilities to progress the works. Particularly, in works package contracts and sub-contracts, the right to alter the manner and sequence of working, or to release works other than in accordance with the sub-contractor's preferred order, is often specifically reserved by the respondent. It follows from Chapter 3 that in the main such clauses mean what they say and are difficult to strike down. The respondent may well be able to argue that the claimant's tender took no or no proper account of the fact that his work was to be subject to a particular restriction and thus allowance should have been made for the fact that output was going to be less than the optimum. This point can be seen clearly in the decisions in *Martin Grant Limited* v. *Sir Lindsay Parkinson & Co Ltd* (1984) 29 BLR 31 and *Kitsons Sheet Metal Limited* v. *Matthew Hall Mechanical and Electrical Engineers Ltd* (1989) 47 BLR 82.[8.4]

The argument that the tender assumed wrongly that optimum output levels would be achievable is closely linked to arguments concerning the programming of the works. These arise in a number of situations:

- The tender assumption that the works will be capable of execution in accordance with a particular programme whereas in fact the respondent reserves the right to alter the order or sequence of the works.
- The assumption that the works can be executed in the order which suits the claimant and thus allows him to optimise productivity whereas in fact the contract allows the respondent to impose a programme of his choosing upon the claimant

Alternatively it may simply be that the claimant fails in his tender to take account of the way in which output levels may be affected by the variables listed earlier.

To the extent that the adequacy of the tender can be attacked, does this mean that the whole claim fails? The argument is that, because the works could not be built using the resource allocated by the tender, any attempt to use this as a base point from which claims for additional time and money can be made is unrealistic and should therefore be disallowed. There is no English authority on this point and the cases decided by the American Boards of Contract Appeals are of limited help because (and of course) all are dependent on their individual facts. The answer is probably to say that it will be a question of degree to be decided on a case by case basis. In the most extreme cases it may be possible to say that the tender bears so little relation to the actual work content of the job that any attempt to measure output by reference to it will be impossible. In the majority of cases, however, it will be possible to ascertain the levels of resource which *should* have been allocated and start from this as a reference point rather than the actual tender.

The more frequently encountered problem is that of determining what the tender actually amounts to. This may be because the tender is insufficiently detailed, perhaps because the records to demonstrate its build up do not exist (either because they have been lost or because they were never created). Alternatively, it is not uncommon to encounter the situation where the tender represents little more than an educated guess, particularly in works which cannot easily be assessed or measured by conventional methods.[8.5] From a practical perspective the claimant's task is then to show that the actual levels of resource employed could be obtained in a way which allowed the works to be completed for the price. This may be more complex than simply showing that the price was sufficient to accommodate the necessary levels of resource. Frequently it will also involve showing that it was possible to obtain particular quantities of scarce resources. This is often an issue in particularly complex installations which need the deployment of personnel with particular skills of whom there may only be a few available in the country.

Attempting to draw these various threads together Fig. 8.3 comprises a simple flow chart indicating the activities to be undertaken and the potential problems experienced in relation to a possible claim by a trade contractor against a main contractor.

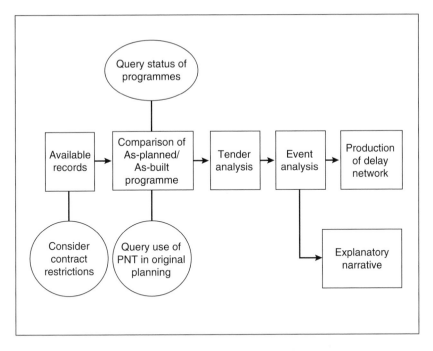

Fig. 8.3 Flow chart of a possible claim.

8.3 *Cause and effect*

Once the claimant has satisfied himself that he could have carried out the works for the tender sum and that the allowed resource levels would have been sufficient to meet any programme or other sequencing restraints, it is appropriate to move on to the question of why the works were actually delayed.

Comparative analysis

The starting point is the production of an 'as-built-programme'. As the name suggests this is a retrospective exercise showing the time periods *actually* occupied by the construction of the works, to distinguish it from the 'as-planned-programme' or any other programme produced for the purpose of planning the works or monitoring progress as the works procede. Whichever method of analysis is adopted, it will depend upon identifying the periods of time actually occupied by the activities comprising the works. Unless it is possible to state with precision what happened and how long it took, almost any subsequent attempt at analysis is likely to be futile. While it may be possible to make an informed judgement as to the effect of delaying events based on the likely effect of particular events judged against the tender output levels, such an approach is always going to be a distant second best, and is always subject to the criticism that it does not actually demonstrate what happened.

At the outset the as-built-programme is best represented in bar chart form. This will not demonstrate the inter-relationship between the delays and their consequences. The object of the analysis of the delays is to show their effect on the programme and the delays which result. If the works have been planned using the sort of project network techniques described in Chapter 7 it may be possible to produce a contemporaneous record of the delays as they progress. This is certainly a helpful start but may be misleading, not least because the true delaying effect of particular events may not become apparent until those consequences actually manifest themselves.

A simple example of the comparison between the as-planned-programme (that which the contractor intended to construct) and the as-built-programme (showing what he actually did build) is shown in Fig. 8.4. This illustration deals with the construction of a wall. Here it will be noted that the delays can be attributed to two causes – correction of defective foundation work by the contractor and the addition of special brickwork detailing. We may assume that the former does not entitle the contractor to an extension of time but that the latter does.

Impacted-as-planned technique

To keep matters simple, the two delaying factors are entirely separate. The causes of delay are also quite clear. In this situation it is unnecessary to analyse matters further. There is no difficulty in identifying the matters which will give rise to an extension of time and no difficulty in determining what that extension should be. This approach is known as the impacted' or 'adjusted-as-planned' technique. The simplicity of this approach also means that it is only

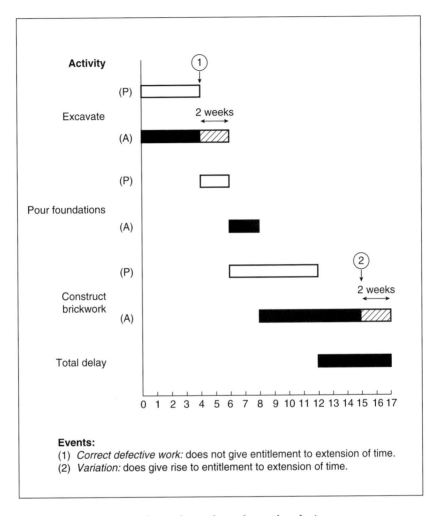

Fig. 8.4 The impacted or adjusted as-planned technique.

useful in the simplest cases. Figure 8.5 shows a situation only slightly more complex than that in Fig 8.4. Again there are two delaying factors in the construction of the wall. In this instance, the delays occur again because of the addition of brick specials as part of the end detailing but also because of a change in the brick source. The first thing to note is that the delays overlap in part (Fig. 8.5(1)). Necessarily it becomes impossible to tell whether the delay can be said to be due to one cause or the other and whether one led to the other, and if so to what extent. Consideration of which delay was

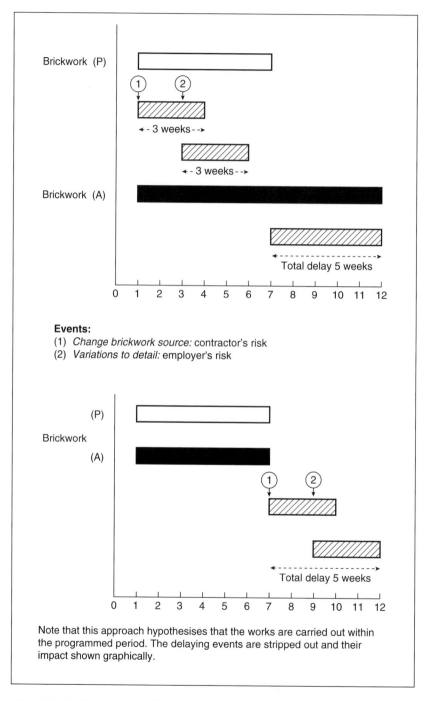

Brickwork (P)

① ②

←- 3 weeks -→

←- 3 weeks -→

Brickwork (A)

Total delay 5 weeks

0 1 2 3 4 5 6 7 8 9 10 11 12

Events:
(1) *Change brickwork source:* contractor's risk
(2) *Variations to detail:* employer's risk

(P)

Brickwork

(A)

① ②

Total delay 5 weeks

0 1 2 3 4 5 6 7 8 9 10 11 12

Note that this approach hypothesises that the works are carried out within
the programmed period. The delaying events are stripped out and their
impact shown graphically.

Fig. 8.5 Collapsed as-built analysis.

actually critical to the progress of the works becomes necessary. In turn this requires a degree of evaluation of the delays which the simple as impacted programme does not permit.

Of course, this may well be unimportant if both delays can be attributed to the employer. It is sufficient to say that events occurred which caused particular delays. However, in Fig. 8.5, while the addition of the brick specials is almost certainly a variation instructed by the employer, the change in supplier may have many causes which may include any of the following:

- instruction by the employer
- design change ordered by the architect
- unforeseen non-availability of materials
- inadequate selection of materials by the contractor
- default on the part of a domestic supplier.

Of these, the first and second are likely to be the responsibility of the employer, thus entitling the contractor to an extension; the third may, depending upon whether the usual alteration to clause 25.4.10.2. of JCT 98 has been made; the fourth and fifth are likely to be the responsibility of the contractor.

Two further problems with this approach are that it assumes that the works were actually built by reference to the programme and secondly that it does not show how the works may have been disrupted. If these matters are prevalent in the works which are being analysed, it is likely that the impacted as-built approach will be discarded. In relatively simple projects where the constituent parts of the project are easily separated, it is a useful approach because of its simplicity, but the shortcomings of the technique will become apparent as soon as the works become more complex. Although what it does is to highlight the crucial importance in any analysis of the availability of records showing the events which occurred and the delays. This must be the precursor to any subsequent analysis of how these delays interacted and how they may have delayed the works as a whole.

Collapsed as-built technique

Where the impacted-as-built technique is not appropriate, a more sophisticated approach is called for. The starting point remains the accurate identification of facts. What is then required is the ability to sift and weigh the relative impact of differing causes of delay. This is known as the 'collapsed as-built' technique. As stated, Fig. 8.5(1) shows the as-built-programme for the construction of a wall, and the delays have been highlighted. Both of the matters identified above have caused part of the delay (and we will assume the deletion of clause 25.4.10.2. from JCT 98). Fig. 8.5(2) then repeats the programme abstracting the delays and displaying them at the end of the last planned activity. This shows the length of delay which is attributable to events which give rise to an entitlement to extensions of time (divided as appropriate between those entitling the contractor to compensation for delays and those which do not); and those which are the contractor's responsibility and do not give rise to any entitlement to an extension of time.

The two major advantages of this approach can be identified quite easily:

- Provided that the records are sufficiently comprehensive it necessarily involves a detailed consideration of the matters which have occasioned delay. It is therefore a useful tool for identifying which factual matters are capable of agreement without debate and which are contested.
- This will include not only the matters which may give rise to an extension of time, but will also require the contractor to identify those matters which were caused by his own default.

However, in the majority of cases it will be no more than a convenient technique for listing the delays and differentiating those which give rise to an entitlement to an extension from those which do not. What it does not do is prioritise those delays enabling the claimant to work out which actually impacted critically upon his works. Particularly it fails to consider the inter-relationship between delays and takes no account of the effect of float. For example, in Fig. 8.5, if inadequate selection of materials has led to a design change which in turn has led to the non-availability of materials, the result is wholly misleading.

Time impact analysis

The solution to this problem is the use of what is called 'time impact analysis'. This approach has been adopted in numerous US disputes. Once again the process requires the records to demonstrate the events which occurred in sufficient detail to allow the claimant to identify each delaying event as a separate entity. Fig. 8.6 applies this approach to the construction of a wall. The as-planned-programme shows that an instruction was issued which added to the duration of the foundation construction. The duration added by this instruction is then added to the activity duration for the foundations. The programme is then re-drawn to show the effect of that delaying event on the works as constructed. The difference between the two will be the delay to completion of the works caused by that event.

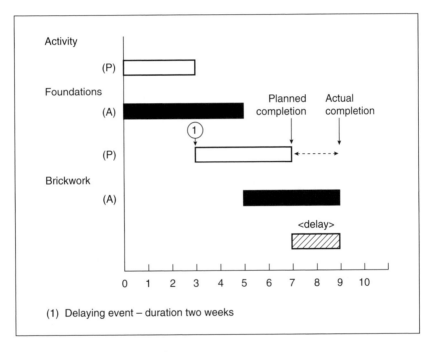

Fig. 8.6 Time impact analysis.

The advantage of this approach is that it is capable of being used during the currency of the works themselves, and can show not only how delays have affected completion of the works but how the delays, once identified, will affect the following activities.

In theory there is no reason why this programme cannot be updated daily, or at any greater or lesser interval which shows the impact of delays or changes *as they occur*. Naturally this allows the production of an analysis which is immensely detailed, but, more importantly, it allows the production of a completely factual analysis of the delays and their consequences. The true strength of this approach is that it is very difficult to attack. The claimant can show that:

- the programme allowed for events to occur in a particular order, and this was possible, and
- events occurred which impacted on the programmed sequence of events in a particular way, and
- the resultant delays can be explained directly by reference to those events.

Therefore, unless the respondent can show that these events did not occur or there is some entirely independent explanation for the delays, the claimant's position will be a strong one.

The disadvantage of this method is the need for complete records to be used to create an 'as it happened' view of the works. It will be recalled that in *McAlpine Humberoak* v. *MacDermott* (1992) 58 BLR 1 Lord Justice Lloyd said in a frequently quoted passage:

'The judge ... dismissed [the defendant's] approach to the case as being "a retrospective and dissectional reconstruction" by expert evidence almost day by day, drawing by drawing, TQ [technical query] by TQ and weld procedure by weld procedure, designed to show that the spate of drawings which descended on [the plaintiff] virtually from the start of the work really had little retarding or disruptive effect on its progress. In our view [the defendant's] approach is just what the case required.'

The only complaint which might be made about this clear statement is that, while the analysis was undoubtedly undertaken by the defendant's expert, it was no more than a particularly thorough assembly of factual material. In the vast majority of instances the records to allow this level of detail are not available. Where it is available this approach almost inevitably produces an extremely complex analysis and care is required in the presentation of this sort of claim to ensure that it is presented digestibly.

Snapshot analysis

The fact that sufficiently detailed records do not exist to enable a complete analysis of the delays need not be a reason to reject this approach altogether. Many claims are put forward using what is referred to as 'snapshot' analysis. As the name suggests, individual delays and their effects are identified and analysed. In some instances, the delays to the works can be attributed to particular matters and it is sufficient to identify these matters in isolation, without the need to look at the works in their entirety. In others the delays are approached from the premise that the claimant asserts that the delays which have been identified are indicative of the delays as a whole and that the analysis of part of the works can be applied generally. Frequently this approach disguises the fact that while the claimant may be able to identify some of the delays the records are not sufficient to permit a comprehensive analysis.

The weakness of this approach is that, if the other party is able to point to some matter which is not covered by the snapshot analysis and which might be said to have caused the delays attributed to the matters covered by the snapshot, the value of the snapshot analysis is greatly diminished. This is a particular problem where the claimant has sought to argue that the delays identified in the snapshot are generally applicable. Equally, this approach is vulnerable to the criticism that the claimant has been partisan in its selection of delaying events.

8.4   Analysing the effect of delays

The previous section proceeds on the assumption that the claimant can actually identify the delays and their effects. In most projects this assumption pre-supposes that the claimant can point to the actual events which have occurred and thus to their duration. The as-built-programme is therefore no more than a factual record.

In many cases this will be all that is required. However, this approach will not really do where questions arise as to:

- the concurrency of particular activities
- the inter-dependence of activities, and
- most importantly float.

In other words, where disputes exist as to whether particular delays were actually critical to the progress of the works.

Computer analysis

Keith Pickavance states that this demonstrates that where the argument concerns criticality of particular delays or the inter-dependence of particular activities, proper analysis is only possible using computerised critical path analysis packages. This is an overstatement. It is, as will be seen, the case that in many cases the availability of suitable software is an essential tool, but it is an inescapable fact that the results obtained will depend on the quality of the data which is input. This can be seen in two ways:

(1) Firstly, it depends on the original planning logic. If the works have been planned in a way which miscalculates the resources required to complete the works, whether in terms of the labour or materials required or the time spans necessary in order to accomplish the activities, the software will almost inevitably produce a distorted and unhelpful result.
(2) Secondly, it depends upon the integrity of the data which is input in relation to delays. The mere fact that a claim has been produced using a state-of-the-art package will seldom be proof against an attempt to distort the results to produce a claim particularly favourable to the party putting it forward.

In short, the use of computerised delay analysis is not an end in itself. Where the works have originally been programmed using a software package, as will be seen below, the task of analysing delay can often be accomplished both quickly and effectively. However, where the works have not been programmed using a software package, or where the software used to analyse the delays uses a different logic to that used to programme the works, the first task facing the claimant is that of proving the logic used to identify the effect of the delays. In that respect, it is no different from the task facing the claimant seeking to identify the delays and their effect by non-computer-aided methods.

A sample analysis

Naturally, the starting point is the basis on which the works were originally planned. Fig. 8.7(1) shows part of a programme taken from an actual project which comprises the refurbishment of a large house. Fig. 8.7(2) shows that programme translated into a network. Fig. 8.7(3) then provides a comparison between the actual planned durations of the activities comprising the works and the actual durations showing planned and actual start and finish times and the additional time taken by reference to contract weeks. From this a programme showing as-built activity durations can be produced and Fig. 8.7(4) provides an overlay of the as-planned and actual activity durations. Finally, Fig. 8.7(5) translates this into network form.

In this example, the critical path for the programmed works takes an obvious and logical route. The delays have the effect of delaying certain of the critical durations, which naturally have the effect of delaying the completion date. The critical events which are not of themselves delayed are not affected by the delayed events save that their start and finish dates are delayed proportionately. The non-

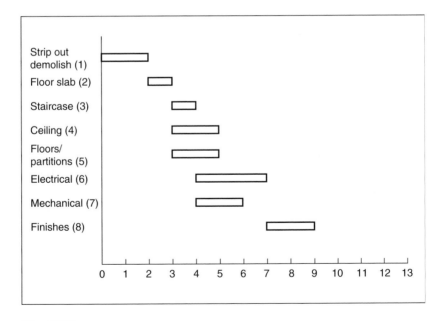

Fig. 8.7(1)

177

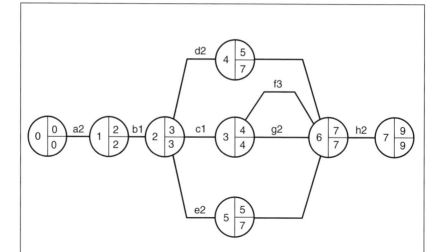

0 = Commence strip out. LFT and EET both = 0
a = period to undertake strip out – 2 weeks

1 = Complete strip out. Commence slab.
EET = 2 i.e. can start at beginning of week 2.
b = pour slab. Duration 1 week

2 = Complete slab. Commence stair, floor and ceiling
c = build stair – 1 week
d = build ceiling – 2 weeks
e = floors – 2 weeks
EET = 3

3 = Complete stair commence M and E works
f = instal electrics – 3 weeks
g = instal mech services – 2 weeks
EET = 4

4 and 5 = Complete ceilings and floors.
Note: while programmed to occur in week 5, this would happen any time up to week 7.

6 = Complete M and E.
Note: although programmed to complete at end of week 6, mech work can complete any time before end week 7.

7 = Finish project – week 9.

Fig. 8.7(2)

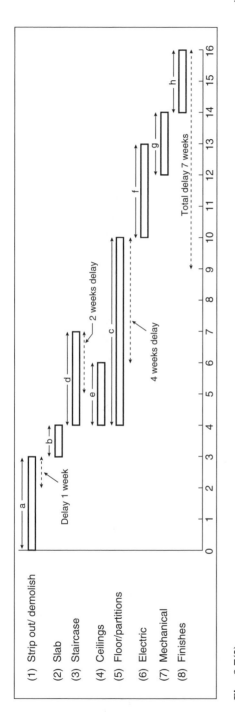

Fig. 8.7(3)

(1) Strip out/ demolish
(2) Slab
(3) Staircase
(4) Ceilings
(5) Floor/partitions
(6) Electric
(7) Mechanical
(8) Finishes

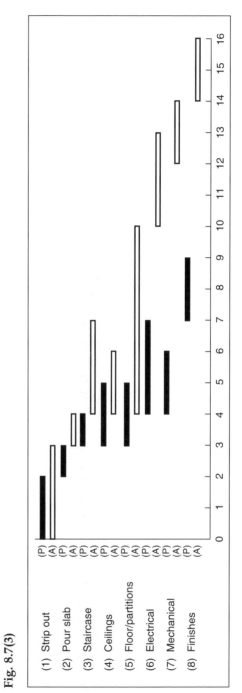

Fig. 8.7(4)

(1) Strip out
(2) Pour slab
(3) Staircase
(4) Ceilings
(5) Floor/partitions
(6) Electrical
(7) Mechanical
(8) Finishes

179

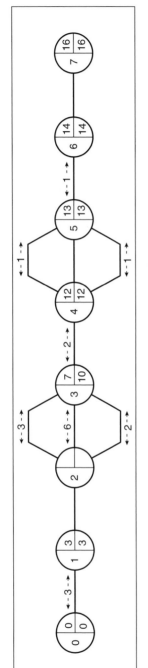

Fig. 8.7(5)

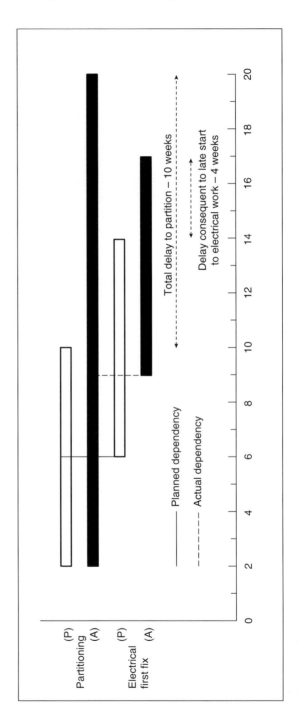

Fig. 8.8 Concurrency.

critical activities also suffer delays in some instances although these have no effect upon the completion of the works.

The delays to the works can therefore be identified simply. The only areas for debate may be in the causes of delay rather than the delays themselves.

Consideration of the as-planned network in Fig. 8.7(2) and the as-built network at Fig. 8.7(5) shows the effect of delays to the concurrent activities – 3, 4 and 5. Applying the approach derived from Chapter 7 it can be concluded that any or all of these programmed events may be critical to the completion of the works. Assuming that no considerations of float arise (which will be addressed below), although Activities 3 ,4 and 5 have equal precedence in the original programme, the effect of delays to Activity 3 causes us to reconsider whether these delays will have any effect upon following activities, and if so what.

Again, the result is (and should be) entirely factual. Three possibilities exist:

- In fact the delay to Activity 3 had no effect on following trades. Plainly in this case Activity 3 could not in truth be said to be critical to completion of the works.
- Although the following trades were delayed that delay did not equate to the whole of the period of delay to Activity 3. It is easy to think of a host of activities where the preceding trades only have to reach a particular point in order to allow a worthwhile start to be made on following activities.
- The delays to Activity 3 caused following trades to be delayed until the completion of Activity 3.

This shows that there is no substitute for the preparation of sufficiently detailed factual records to enable this sort of distinction to be drawn. The description of this type of exercise in *McAlpine* v. *MacDermott* as a 'retrospective and dissectional reconstruction' is particularly apt.

No effect on following trades

Importantly the exercise shows that the claimant may well be able to point to a number of delays which did not actually cause any delay to the works. There is an understandable temptation to regard the identification of delays as an end in itself without addressing their impact on the completion date. Nevertheless, it is entirely possible that to avoid the delays to the activities other than Activity 3 may require the deployment of additional or simply different labour or material resources. The mere fact that delays have occurred does not of itself give rise to an extension of time although this may entitle the claimant to additional payment (under clause 26 of JCT 98 for example) to compensate for the extra labour required.

Re-programming – concurrency

The harder question concerns the situation where the delays meant that the start of the following activities were delayed but not by the whole of the period of delay. Fig. 8.8 shows an example of this showing the critical path for the works – the critical point coming not at the point where the partitioning is completed but the point at which sufficient of the partitioning has been completed to enable the first fix electrical works to be progressed. The issue here is the inevitable debate as to the point at which the following activity could or should have been commenced. In summary:

- The partitioning has been delayed by a total of 10 weeks.
- The first fix electrical trunking clearly depends upon completion of enough of the partitioning to enable the electrical work to follow on behind.
- As originally programmed the electrical first fix works were scheduled to commence 4 weeks after the commencement of the partitioning. However, the delay to the partitioning meant that after 4 weeks the partitioning was insufficiently advanced to enable a meaningful start to be made.
- In fact the electrical works could not be commenced until the end of week 9. The question requiring further consideration is whether the delay of 5 weeks can actually be attributed to the lack of progress to the partitioning or some other matter not attributable to the respondent.

Rarely will the available records be of such detail that the claimant can produce an accurate comparison between the precise position of the partitioning at the end of each week from the end of week 4 onwards to see precisely when the works had actually reached the point at which it was contemplated that the electrical trunking could be commenced. It is rare, however, that the argument that the claimant should, for example, have commenced the electrical trunking before he actually did, will much assist a respondent. Firstly, where the delay has occurred due to some default on the part of the respondent, he will have difficulty in persuading any third party that the 'wronged' party should have done other than he did, except in the grossest circumstances. Secondly, where the facts demonstrate that works programmed to start on a particular date were delayed by reason of a matter entitling the contractor to an extension of time, then, unless the employer can point to some form of evidence, he will not be able to substantiate a claim that the period of delay should have been less than in fact it was.

While it is therefore customary to see provisions in the preliminaries to contracts which provide that the contractor will do everything reasonably practical to minimise the effects of any delays, the burden of proving that the contractor failed to minimise the delay will rest with the employer.

A related consideration is the right to instruct the delay to be diminished by the acceleration of the works. The majority of standard forms do not provide an entitlement to accelerate the works, but it is increasingly common to see tailor-made contracts providing the employer with the right to require acceleration. The effect of any provision entitling the employer to accelerate will depend upon the individual circumstances. What is clear is that while the effect of such an instruction is to reduce the effect of the delay, it may entitle the contractor to recover the costs of the acceleration.

In both scenarios, it is also clear that the issue is what actually occurred. It is not appropriate for the contractor to produce a claim for an extension of time which is advanced on the premise:

- 'Here is the extension of time to which we would have been entitled but for our measures to mitigate the effect of the delays' or
- 'Our extension of time is calculated by reason of the delaying events aggregated together and discounting the effect of the measures we took to accelerate.'

183

although these approaches are not uncommon. The fallacy of this sort of claim is that it seeks to produce a theoretical entitlement to an extension of time unrelated to the events which actually occurred, and even less related to the effect of one event upon another.[8.7]

Critical delay

The final possibility identified above is that where the delay to Activity 3 prevents any progress being made on following activities until the completion of Activity 3. In other words, Activity 3 is truly critical to the progress of the works. This will frequently be the case where the activity in question leads to a particular milestone in the works such as the achievement of weather-tightness or 'power on'.

8.5   Float

The third situation described above will occur only where there is no float to absorb the effect of the delays.

This leads to the debate as to what happens where delays occur which cause programmed float to be absorbed in whole or part without necessarily causing any delay to the works or to the following activities. Fig. 8.9 demonstrates this – Activity 1 is programmed to take six weeks but will actually only occupy two of these weeks. In fact the whole of that period is used but no actual delay is caused to activity 2. The question 'Who owns the float?' is frequently asked. Applied to the situation shown in Fig. 8.9 what this really means is – does any consequence flow from the absorption of the float, either in terms of time or compensation?

This issue has been considered at length by a number of commentators and authors without any clear consensus emerging. Some have suggested that the float time necessarily 'belongs' to the project and that therefore whichever party comes to utilise the float first should have the benefit of it; while the remainder contend that float is a function of the contractor's planning process and that therefore he builds the float into his plans to hedge against the possibility that he might take longer to undertake the various activities than he had allowed and thus requires a margin for error

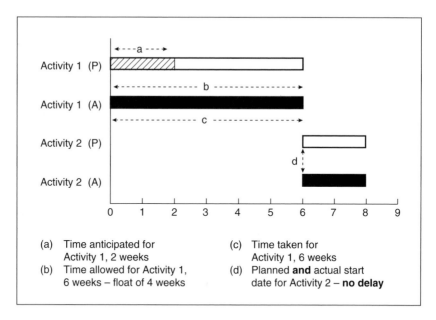

Fig. 8.9 Who owns the float?

which should be maintained by the grant of an extention of time.[8.8]

In fact consideration of the role of float from first principles shows that the debate is less complex than might be supposed.

(1) In the majority of standard form contracts, the programme is not a contract document. The contractor's obligation is to carry out and complete the works by the completion date, rather than by any specific activity date.

(2) Accordingly, unless the effect of delaying a particular activity is to cause delay to the completion date of the works, the programme is to be regarded as a planning tool and no more.

(3) Within the constraints of the need to complete the works by the date for completion, the contractor can programme the works as he wishes.

(4) Similarly, if the employer's conduct causes the contractor to use up some or all of the float without causing delay to the works, the consequence may be disruption if the contractor can identify the need to deploy additional resource, but it will not entitle him to any extension of time.

The point is the same as that considered in *Glenlion Construction* v. *The Guinness Trust* (1987) 30 BLR 89. It will be recalled (see Section 4.3) that the claimant had produced a programme showing completion taking place prior to the contract date for completion. Judge Fox-Andrews held that the claimant was certainly entitled to complete by the date shown in the programme but was not entitled to the incorporation of a term obliging the employer to provide information in sufficient time to allow the contractor to complete by the programmed date.

The effect of the decision (which, it is suggested, is clearly sensible) is that float belongs to neither party and is to be regarded as a neutral commodity which exists for the benefit of proper planning of the works. This illustrates that in many projects, particularly those which are resource driven (that is to say where the time allowed for an activity substantially exceeds the work content in order to allow the contractor flexibility to carry out tasks in the order which best suits the requirements of the works) the contractor will have to demonstrate delays which absorb the whole of that float in order to show further delays entitling him to claim extensions of time.

The more important consideration is that the use of float may have the effect of altering the critical path of the works. In Chapter 7 it was seen how the critical path for the works is derived from the network produced for the works and that the critical path will run through the activities with zero float. Where the works are delayed and float is used up, the effect will be to make activities which were not previously critical, become critical. This is illustrated by Fig. 8.10 which again derives from the refurbishment works example seen above. The decorative finishes (Activity 25) are not originally programmed to be a critical activity and are shown as having a large element of float. However, the effect of the cumulative delays to the partitioning and first and second fix electrical works is to absorb the whole of the remaining float such that the available time for these works is equal to the time required to carry out this activity. Any delays to the decorations will have the effect of delaying the completion date. The decoration therefore becomes critical to the completion of the works. If the effect of the delays to the preceding activities had been that the time available for the decorative finishes was less than that programmed then the completion date for the works would have been delayed. It would then be possible to say

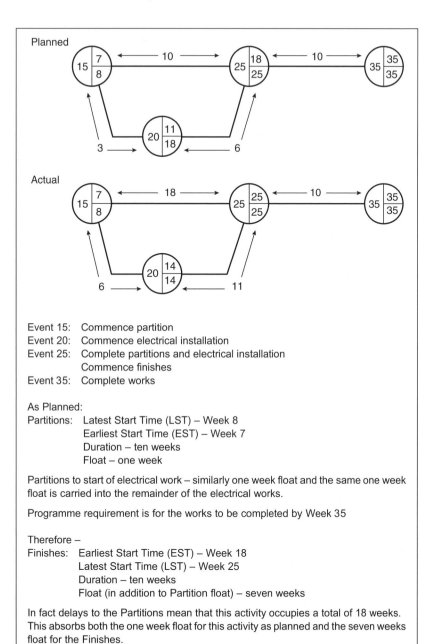

Planned

Actual

Event 15: Commence partition
Event 20: Commence electrical installation
Event 25: Complete partitions and electrical installation
Commence finishes
Event 35: Complete works

As Planned:
Partitions: Latest Start Time (LST) – Week 8
Earliest Start Time (EST) – Week 7
Duration – ten weeks
Float – one week

Partitions to start of electrical work – similarly one week float and the same one week float is carried into the remainder of the electrical works.

Programme requirement is for the works to be completed by Week 35

Therefore –
Finishes: Earliest Start Time (EST) – Week 18
Latest Start Time (LST) – Week 25
Duration – ten weeks
Float (in addition to Partition float) – seven weeks

In fact delays to the Partitions mean that this activity occupies a total of 18 weeks. This absorbs both the one week float for this activity as planned and the seven weeks float for the Finishes.

Thus the Finishes Earliest Start Time (EST) becomes Week 25 and hence there is no available float.

Fig. 8.10 Where the float is already used up, non-critical activities (here, decorative finishes) become critical.

that this activity therefore had negative float. In simple terms, the situation where the time available to carry out the works is less than that actually required to execute it will mean that the project is in delay.

This also demonstrates a problem which can be encountered. Activities will only become delayed when the float for those activities has been absorbed. Where that delay has consumed all of the available float for the activity in question and has eaten into the float for succeeding activities, resulting in negative float, the following activities will be delayed. It follows that a network based approach will only serve to explain the cause of delays, and will not distinguish whether those delays result from events of contractor delay or employer delay. Having isolated the causes of delay, responsibility for those causes can be identified by use of time impact analysis discussed above (pp 172).

CHAPTER NINE
PRESENTATION OF THE CLAIM

9.1 *General*

There is no such thing as the ideal format for a delay claim. We have already seen that each claim is dependent upon the individual facts of its own case and the purpose for which it is produced. The considerations governing a claim which is intended to form part of formal legal proceedings, whether in court or arbitration, will be wholly different from those which apply where the claim is merely intended to form an opening salvo in negotiations.

Universal guidelines

There are, however, some simple guidelines which are applicable to almost all claim submissions:

Summary

It is often a useful exercise to start the submission by summarising exactly what the claim needs to demonstrate, for example that

- particular activities were delayed
- by a specific period of time
- which caused (or is likely to cause)
- an identifiable delay to the project.

That can usefully be followed by a statement of the circumstances in which the claim was produced, for example that the works are in progress, that particular delays have occurred, it is anticipated that those delays will cause specific and identifiable delays to the overall completion of the project, and therefore the submission is made for the purposes of obtaining an extension of time pursuant to, for example, clause 25 of JCT 98.[9.1]

End user

It is crucial to bear in mind the end user of the claim. Put another way, what does the reader of this claim need to be told in order to make the desired decision? For example, if the claim is aimed at the architect or engineer who has administered the project, it may be safe to conclude that he will not need extensive background information concerning events leading up to but with no direct bearing on the matters in issue. By contrast, if the end user is a judge or arbitrator it is highly likely that he will need at least some understanding of the project because he has had no prior involvement in it.

Simplicity

Simplicity is always a virtue. Even in the most complex claims, the object should be to present the material in a way which the reader will readily understand. If a particular proposition is relied upon, it is also sensible to explain why. It is inherently dangerous to allow facts or assertions to speak for themselves as the risk is that they will be interpreted in a way which is not intended.

Facts

Facts are the best means of persuasion. There are a few prizes for rhetoric in construction claims. In claims which are produced for negotiation purposes prior to the commencement of proceedings, emotional or argumentative language is likely to be counter-productive; invariably, references to 'flagrant breaches' or 'blatant disregard for contractual obligations' will be interpreted as a criticism of the very person the claim seeks to persuade. If the claim is intended for a judge or arbitrator, it is unlikely that he will be impressed.

The remainder of this chapter therefore looks at how a claim submission might be put together. Consideration is also given to the part to be played by expert witnesses and witnesses of fact, and consideration is given to the use of particular presentation methods, notably the Scott Schedule.

9.2   *Putting together the submission*

The example chosen concerns a claim by a roofing and cladding sub-contractor engaged by a main contractor as part of a substantial project. The example is not based on any particular case but borrows details from a number of actual disputes.

The works themselves are to be carried out in accordance with a form of sub-contract comprising an extensively amended version of DOM1. The purpose of the amendments is clearly to attempt to shift the burden of risk as far as possible onto the sub-contractor. The works package comprises a number of identifiable activities and locations. The total delay amounts to 25 weeks in addition to a contract period of 40 weeks.

In particular, the sub-contract has been amended to provide that the main contractor shall be entitled to require the sub-contractors to co-ordinate the works in such a way as to ensure progress of the main contract works. The sub-contract also provides that the sub-contractor shall carry out and complete the works by the date stated in the Appendix. The sub-contractor's right to claim an extension of time is regulated by the entirely typical clause providing that where it becomes apparent to the sub-contractor that the progress of the works is likely to be delayed, he shall give notice to the main contractor of the fact of the delay and the circumstances giving rise to the delay, providing his best estimate of the likely extent of the delay and its effect upon other matters.

During the initial weeks of the project, it is clear that the works on the first area, Zone 1, underwent significant delays caused by late release of steelwork. (See Fig. 9.1 p. 194.)

It is important to remember that the primary purpose of the submission is to inform. Elaborate or argumentative language should be avoided where possible. Most claims in these circumstances deal with specific events and the claim should aim to be a snapshot capturing those events. While the format will depend upon individual circumstances, the following approach is often helpful:

- a brief introductory statement identifying precisely what the submission seeks to achieve, e.g. a particular extension of time;
- a brief statement of the particular contract terms on which reliance is to be placed;
- a summary of the events which it is contended caused the delay;
- a summary of the consequences of the delaying events;
- identification of the notices served and relied upon.

In the main, the temptation to annex large numbers of documents is to be resisted. In particular, quantities of correspondence will seldom serve much purpose, especially where that correspondence serves primarily to advance arguments rather than to identify particular events. The conspicuous exception to this is that photographs are often an extremely persuasive tool, particularly if dated.

The next stage is often the submission made to assist the architect in his review pursuant to clause 25.3.3 or the engineer in reaching a clause 66 decision. While the original claim submission is intended primarily to inform, it goes without saying that by the time of the architect's review or the clause 66 decision, the architect or engineer has already given indications of his view which are not to the liking of the contractor. The nature of construction disputes mean that there is seldom much doubt as to the reasons why the initial submission has been rejected. The rejection of claims generally occurs for one of two reasons:

- the lack of information which is said to be crucial to enable to architect or engineer to make his decision; and/or
- disagreement with the reasons given in support of the application for an extension of time.

In the former case, the claimant's task again is to inform. In most instances, it benefits the claimant much more to provide this information, wherever possible, rather than argue that the respondent is not entitled to it or does not need it. To state the obvious, if the architect or engineer says that he requires a particular piece of information or information in a particular form in order to make his decision, if that information is provided in the form he requires, it will be significantly harder for him to resist the claim.

It is sometimes suggested in these circumstances that, if the architect or engineer indicates that he requires a claimant to provide detailed particulars of cause and effect equivalent to that which would be required if the matter proceeded to an arbitration or litigation, the claimant should not have to produce that degree of detail. This is a difficult argument. The practical answer is probably that in these circumstances that degree of detail should be provided at this stage because, without that degree of particularity, it appears inevitable that the architect or engineer will continue in his view, and proceedings, requiring the claimant to particularise his claim with that degree of precision anyway, become more likely.[9.2]

Where the architect or engineer has rejected the initial claim stating that he disagrees with the reasons for seeking an extension of time, it will frequently be difficult to cause a significant rethink at the time of the clause 25.3.3 or clause 66 review. However:

- it is always worth asking whether rejection of the initial claim has occurred simply because of the way in which it was presented;
- the review process must be regarded as an opportunity to consider the wider picture.

In particular, it is not uncommon to find that the decision to reject the initial claim has occurred because the architect or engineer has attempted to consider matters which were not raised in the initial submission.

The initial claim submission in Fig. 9.1 is based on the premise of delays occurring to Zone 1. This occurs by reason of late release of steelwork. Although programmed to commence in Week 1, late release of steelwork prevented this work from commencing until Week 5, at which point the works should have been completed and the sub-contractor should have been able to move on to Zone 2. The initial claim submission, produced during Week 5, concentrates understandably on the effects on Zone 1. However, at the end of Week 8 when the architect replies to the submission, there is no doubt that Zone 2 is in delay. Nor is there any doubt that the architect is right in saying that the sub-contractor failed to have sufficient manpower available to progress works in Zone 2. In this instance, of course, the architect's error is not to realise the sequential effects of delays in Zone 1 on the works in Zone 2.

The form of submission at this stage will obviously depend entirely upon the particular circumstances. However, a development of the basic approach outlined above may well prove useful:

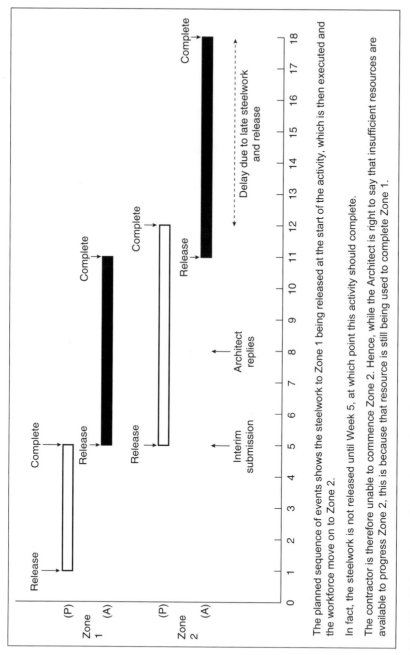

The planned sequence of events shows the steelwork to Zone 1 being released at the start of the activity, which is then executed and the workforce move on to Zone 2.

In fact, the steelwork is not released until Week 5, at which point this activity should complete.

The contractor is therefore unable to commence Zone 2. Hence, while the Architect is right to say that insufficient resources are available to progress Zone 2, this is because that resource is still being used to complete Zone 1.

Fig. 9.1 Delays due to late release of steelworks.

- Recap on what is sought, identify particularly whether this differs at all from the original submission, and if so why.
- Identify the contractual basis on which this is sought, identifying and dealing with any contractual arguments which have been raised in rejecting the original submission.
- Summarise any other reasons given for rejection of the original submission.
- Identify further information supplied, explaining why this will affect the previous reasoning and, if appropriate, why the previous decision of the architect or engineer is flawed.

Again, brevity is a virtue. The temptation to be argumentative should be resisted.

The question 'When does this become a legal dispute?' has already been considered in Chapter 6. Often, that decision is taken when an initial claim submission, perhaps augmented by a further submission, has nevertheless been rejected. However, almost as frequently, the claimant perceives that his best course of action is to produce a more detailed submission in the belief that this will further his negotiations seeking an extension of time. It follows from what we have seen above that the question – when does this become a legal dispute, can almost be better characterised as – when do we need to decide to produce a fully detailed and substantiated claim. The reverse process is also valid in that the preparation of a fully detailed claim will often be accompanied by the realisation that the claim in question may in due course form the basis of formal pleadings in arbitration or litigation.

It also follows that this decision must be accompanied by a candid appraisal of what can actually be proved and the resolution to devote sufficient resource to the preparation of the claim to enable it to be properly formulated. This is reinforced by the provisions of the new Civil Procedure Rules from which it is clear that, when proceedings are commenced, they need to be fully particularised. The notion of incomplete proceedings in general terms commenced by way of a warning shot now appears to be a thing of the past.

This decision must also be accompanied by a realisation that the function of the claim document is no longer merely to inform or to cajole an architect or engineer but to prove the claimant's case on the balance of the probabilities. In the example given above, the

195

question is therefore whether the sub-contractor can prove that his works suffered a 25 week delay which is attributable to matters entitling him to an extension of time.

The basic methodology set out above is still likely to prove appropriate. However, it is at this point that attention must be paid to questions of causation. The claimant must ask whether he has the data available to produce the sort of detailed analysis of delays and their consequences identified in Chapters 7 and 8, and what this will actually show. Consideration should be given to precisely how the claimant proposes demonstrating the effect of the delays complained of and the particular logic to be applied. In the example given above, it may be that the claimant can demonstrate that in fact the effect of the delays can be analysed very simply by reference to the consequences of late release of steelwork, a very simple path can be plotted through the works identifying the dates when the steelwork should have been released with the dates when it actually was released. This is shown on Fig. 9.2 which also demonstrates the absorption of float due to late releases of steelwork in Zone 1 and the consequent critical nature of the delays in following zones.

It is quite clear that this approach would be more effective than that adopted in many claims which is to produce a serious of lengthy 'claim narratives' seeking to identify all of the events which did cause or might have caused or contributed to delays.

9.3 Presenting the evidence

Scott Schedules

Scott Schedules have long been regarded as synonymous with construction disputes. They were invented by Sir George Scott KC, an Official Referee during the 1920s and 1930s. They are a convenient way of presenting large quantities of detailed evidence. An example of how a Scott Schedule might be used in a delay claim is provided in Fig. 9.3.

Undoubtedly this approach is more easily digested than lengthy claim narratives. However, while Scott Schedules are frequently a useful way of ordering information, they do not lend themselves to demonstrating the interaction between delaying events.

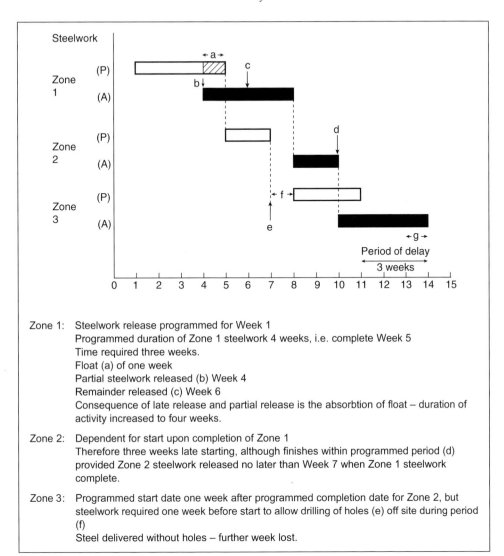

Fig. 9.2 Absorption of float due to late delivery of steelworks.

Serial No.	Activity	Event	Cause	Contract Ref	Consequence	Respondent's Comment	Arbitrator's Comment
(1)	Plumbing 1st fix	Pipework/ Ductwork clash	Drawing error (Ref M/201)	2.4.1, 25.4.6	5 days delay prior to issue of drawing M/ 201(a) (concurrent)		Late access
(2)	Plumbing 1st fix	Lack of access to Plant Room	Delay by preceding trade	25.4.12	10 days until Room released (critical)		

Fig. 9.3   Scott Schedule.

9.4   *Witnesses of fact*

At the time when the decision is taken to produce a fully detailed claim, it is invariably sensible to ask what facts need to be proved in order to sustain a particular assertion, and which witnesses exist who are capable of providing that evidence. It is worth remembering that in many cases the claim will stand or fall according to whether witnesses of fact will give evidence which supports or undermines particular contentions regarding delays. Although delay claims will generally require experts to give evidence as to the method adopted for showing the interaction between delays, that evidence is of little use if the facts on which that analysis is based cannot be proved.

In the past witness statements were frequently prepared shortly before trial. The practice of exchanging witness statements grew up during the mid to late 1980s. Prior to that it was not uncommon for parties to go into a trial with no fixed idea of the evidence to be provided by their opponent's witnesses. The effect of the Civil Procedure Rules is to bring forward this process. In many cases, it will be necessary to serve witness statements with the original claim. While this causes a good deal of work to be done earlier than would otherwise be the case, two benefits are achieved:

- It enables statements to be taken closer in time to the events which those statements describe.
- It encourages both a frank appraisal of the party's own position

and a greater ability to judge the strength of an opponent's position.

Witness statements themselves should ideally be succinct, chronological summaries of the matters of which the witness has first-hand knowledge. Whilst the rules governing hearsay (that is to say matters of which the witness does not have direct knowledge) have been considerably relaxed, self-evidently the weight to be afforded to second-hand evidence will obviously be less than that given to matters within the witness's own knowledge.

In most cases statements will chronologically follow events. It may be convenient to draft a statement by reference to particular claim headings or areas of works. As far as possible, the statement should be in the maker's own words and, when drafting the statement, the witness should be taken to the particular documents to which reference will be made. Rather than quoting at length from the documents, it is often helpful to append a schedule of the relevant documents to the statement. It is worth noting that this is often a more convenient approach than annexing large quantities of documents to the claim itself.

Finally, it is always worth remembering that the witness statement is intended to be a summary of the facts relied upon, rather than an attempt to argue the case. Confrontational or argumentative witness statements invariably act as invitation to the opponent vigorously to cross-examine the witness.

9.5   *Claims consultants*

It has become common over the past 20 years, even in large organisations, for the task of claim preparation to be sub-contracted out to claims consultants. Claims consultants are professional preparers of construction claims although their role is frequently broadened into providing litigation support for solicitors and barristers and on occasions has extended to conducting construction arbitrations.[9.3]

Because claims consultants generally come from a range of construction disciplines and because their exact function is sometimes difficult to define, there can be misunderstanding as to the position of the claims consultant. While the authorities[9.4] make it clear that the claims consultant's costs can, in certain instances, be recovered as part of the costs of the action or arbitration, there is no clear authority as to the precise legal role a claims consultant occupies. For example:

- It is probable that where the claims consultant conducts proceedings without solicitors, he will be unable to claim the benefits of legal professional privilege.[9.5]
- Claims consultants are not subject to any single regulatory body. While this may provide advantages in that it gives greater flexibility over fees and the particular basis on which a claims consultant can charge for his services, it carries obvious disadvantages.

What is clear is that the claims consultant engaged to produce the claim on behalf of his client is not regarded as an expert. He is entitled to put his client's case to its best advantage. His position is really no different to that of the employee of the claimant organisation who is engaged to prepare his employer's case.

9.6 Expert evidence

The position of the claims consultant must therefore be contrasted with that of the expert. It is clear from the cases, particularly *Whitehouse* v. *Jordan* [1981] 1 WLR 246, *University of Warwick* v. *Sir Robert McAlpine* (1988) 42 BLR 1 and *The Ikarian Reefer* [1993] 2 Lloyd's Rep 68, that the duty of the expert is primarily to assist the court. The expert's duty is to present his evidence in a way which is impartial and which presents both sides of the argument. The duties of experts are now set out in Part 35 of the Civil Procedure Rules. Part 35.3 provides:

'(1) It is the duty of an expert to help the court on the matters within his expertise.
(2) This duty overrides any obligation to the person from whom he has received instructions or by whom he is paid.'

The practice directions annexed to Part 35 require at paragraph 1.2(5) that:

'where there is a range of opinion on the matters dealt with in the report [the expert must]:

(i) summarise the range of opinion, and
(ii) give reasons for his own opinion.'

As such, it is almost impossible for the claims consultant who has prepared the claim on behalf of one party subsequently to act as an expert. The point was graphically illustrated in *Cala Homes (South) Ltd* v. *Alfred McAlpine Homes East Ltd* [1995] CILL 1083. The plaintiff's expert, an eminent architect, had his attention drawn to an article he had written some years earlier in the journal of the Chartered Institute of Arbitrators. He had written:

'How should the expert avoid becoming partisan in a process which makes no pretence of determining the truth but seeks only to weigh the persuasive effect of arguments deployed by one adversary or the other?

'[T]he man who works the three card trick is not cheating. Nor does he incur any moral opprobrium when he uses his sleight of hand to deceive the eye of the innocent rustic and to deny him the information he needs for a correct appraisal of what has gone on. The rustic does not have to join in, but if he chooses to, he is fair game ... concealing what is true does indeed suggest what is false, but it is no more than a suggestion just as the three card trick was only a suggestion about the data, not an outright misrepresentation of that. Thus there are phases in the expert's work. In the first he has to be the clients' "candid friend" telling him all the faults in his case. In the second he will, with appropriate subtlety, be what [was] called a "hired gun" so that clients and Counsel when considering the other side's arguments can say, with Marcellus in *Hamlet* "shall I strike at it with my partisan?".'

Mr Justice Laddie did not mince words:

'The whole basis of [the expert's] approach to the drafting of an expert's report is wrong. The function of a Court of Law is to discover the truth relating to the issues before it. In doing that it has to assess the evidence produced by the parties. The Judge is not a rustic who has chosen to play a card of three card trick. He is not fair game. Nor is the truth . . .

'In light of the matters set out above, during the preparation of this judgment I re-read [the expert's] report on the understanding that it was drafted as a partisan with the objective of selling the Defendant's case to the Court and ignoring virtually everything which could harm that objective. I did not find it of significant assistance in deciding the issues.'

Interestingly, Keith Pickavance observes that the role described by the expert in his article was actually that of a claims consultant rather than an expert witness. This view may be a little generous to the expert because it is actually clear that he was seeking to describe the role of an expert and to justify how the role of an expert might be said to change as a dispute evolves. More pertinently, it clearly demonstrates the dilemma which faces many parties – having very properly engaged the services of a specialist to prepare their claim they find themselves obliged to retain another third party to act as an expert. In some cases the consultants seek to remedy this problem by deploying another member of their staff to act as the expert. It goes without saying that this fiction is not really an answer to the problem.

Consequently, the parties frequently face the unpalatable choice between engaging an independent person to act as expert and risking having their original claims consultant's evidence devalued. In fact because the expert and the claims consultant are generally employed to undertake different tasks, the former to prepare the claim, the latter to offer opinion as to its merits, the overlap and duplication in cost will be less than appears first sight.

A partial answer to this problem may lie in the increased use, encouraged by the Civil Procedure Rules of Court appointed experts whose costs are borne jointly by the parties. This practice is still in its infancy and it remains to be seen whether it provides an answer.

In delay claims this dilemma might be less than in other types of dispute. The function of the expert is likely to be limited to an explanation of the methodology used for analysing the delays. This task is likely to need performing irrespective of whether the original claim was produced by a claims consultant or the claimant himself.

CHAPTER TEN
DISPUTE RESOLUTION

10.1   *Changing times*

Both the Latham and Egan reports extol the virtue of creating a conflict free climate for construction.[10.1] Very few construction professionals would argue with this. The same is certainly true of the majority of construction lawyers. All but the most unreconstructed litigator would agree that to spend years and hundreds of thousands of pounds on litigation and arbitration is a poor use of time and resource and that lawyers and consultants would be better deployed in finding ways in which disputes or differences can be dealt with quickly and efficiently in a way which does not hamper completion of the works.

On a superficial level this has led to construction litigators reinventing themselves as 'dispute resolution' lawyers. However, this re-branding is only the start. The defining theme of the past decade has been the energy which has been devoted to devising ways of taking some of the conflict out of construction. This has manifested itself in a number of ways:

- Legislation in the form of the Housing Grants, Construction and Regeneration Act 1996 and the Late Payment of Commercial Debts Act 1998 which have sought to create a fairer climate for doing business with the intention of reducing the scope and number of disputes.
- The growth of partnering as a means to promote conflict free procurement of construction projects together with a commitment to fairer and simpler forms of contract.
- The Woolf Reforms which comprise the most radical overhaul of civil justice since 1875 and seek to simplify the administration of the court system and create a civil justice system which is quicker, fairer and more accessible.

- The reform of the domestic arbitration system by the Arbitration Act 1996 which aims to create a more efficient climate within which to administer domestic arbitration.
- The development of adjudication and ADR as means to resolve disputes without the need for the full rigours of arbitration or litigation.

Two distinct themes emerge. On the one hand is the attempt to find ways in which construction can be administered in a way geared towards the parties sharing goals and working together to achieve those goals. On the other is the recognition that disputes are a facet of commercial life but that when they arise they should be dealt with in a way which does not detract from the project as a whole and allows the dispute to be resolved and the parties to get on with the job in hand.

Partnering

The first of these themes, that of making construction work in a way less prone to creating conflict, lies at the heart of the Latham and Egan reports. It has also led to the growth of project partnering, either on a contract by contract basis or as part of a long-term relationship. Partnering documents generally comprise a set of common goals which the partners intend to adhere to with the object of maximising their mutual gain, and avoiding disputes. The question most frequently asked is whether partnering arrangements (and it is a mistake to call them 'agreements' since they seldom impose legally enforceable obligations on the parties) should operate instead of contract documents or as an addition to them. At the KPMG Partnering Symposium in November 1997 the latter view was espoused by Fiona Hammond of British Airports Authority who said

> 'Was partnering about contracts? Probably not. Contracts were about risk. Relationships were a different matter ... Contracts had become largely empirical, and partnering was about managing that process and building teams. Partnering was about making the contract reflect what the parties had agreed, not what the lawyers thought it ought to mean.'[10.2]

In this respect her views contrast sharply with those of Sir John Egan. In paragraph 69 of his report he calls for

> 'an end to reliance on contracts. Effective partnering does not rest on contracts. Contracts can add significantly to the cost of projects and often add no value to the client. If the relationship between constructors [sic] is soundly based and the parties recognise their mutual interdependence, then formal contract documents should gradually become obsolete. The construction industry may find this revolutionary. So did the motor industry but we have seen non-contractually based relationships between Nissan and its 130 suppliers.'

It would be interesting to judge the response of the industry if the word 'contract' was replaced by, say, 'insurance' or 'health and safety regulations'. While the example given is interesting, it possibly overlooks the fact that in this type of relationship where one party is much more powerful than the other, partnering risks becoming no more than the imposition of terms by one party.

Dispute resolution

The second theme is that of providing quicker and better ways of resolving disputes. In the context of delay claims this could hardly be more important. If traditional methods, litigation or arbitration, are employed the process can and does take years and costs sums which often dwarf the sums in dispute. The development of adjudication and alternative dispute resolution is designed to provide a way of resolving disputes which allows the parties to address the matters in dispute and then get on with the business of being contractors or developers. The emergence of these new methods of dispute resolution coincides with the changes caused by the Arbitration Act 1996 and the Woolf reforms. The defining theme of the Woolf reforms is that the courts exist for the benefit of the parties rather than their lawyers. While this is less explicitly stated in the Arbitration Act, it is apparent that construction arbitrators must regard part of their duty as being to resolve disputes in a manner which is appropriate to the issues and sums at stake.

Like all reforms, criticism has been voiced. Much of it has a common theme, namely that the new forms of dispute resolution (particularly adjudication) and the reforms to litigation and arbitration devalue the process resulting in a lesser quality of justice. That debate is largely one for the jurists but it is worth making a few simple points.

- The assumption that old-style litigation and arbitration mark the ideal forms of dispute resolution is not supported by the facts – the old system was eminently capable of producing perverse and unfair results.
- Given the choice, most litigants would probably prefer a quick and economically affordable result rather than a slow and unaffordable one, even if it meant sacrificing some of the legal niceties. Adjudication meets these objectives.
- Few litigants are really bothered by whether their case raises ground-breaking legal issues.
- Experience suggests that in fact the compromises thought to be inevitable in the wake of the reforms have not been as great as may have been anticipated. Little in the way of legal rigour has actually been sacrificed.

In the context of delay claims – perhaps the most complex species of construction disputes – it might be thought that these reforms would lead to a relaxation of the standard of proof required in order to sustain a claim and thus make the task faced by claimants easier. As yet there is no clear data to support or disprove this. However there are indications that, while the procedural rigours of the dispute resolution process are being eased, the requirement properly to express a claim is not.

The Housing Grants, Construction and Regeneration Act 1996

Anticipated with interest throughout its passage through Parliament, the Act was widely referred to as 'the Latham Act' by those sections of the construction press who believed that it would see the wholesale implementation of *Constructing the Team*. Despite receiving the Royal Assent in 1996 the Act was not brought into force until May 1998.

Sections 104 to 117 of the Act are concerned with construction. For present purposes by far the most significant part of the Act is section 108 which provides a statutory right of adjudication in respect of all disputes arising out of construction contracts entered after 1st May 1998. This is dealt with below.

10.2   Adjudication

The schemes

The right to adjudication provided by section 108 of the 1996 Act is given effect by the Construction Contracts (England and Wales) Regulations 1998 (SI 1998/649). That statutory instrument provides regulations for the conduct of adjudications. Prior to the publication of the scheme in these Regulations ('the Scheme'), the Official Referees Solicitors Association[10.3] had published its own set of rules for adjudication. The two sets of regulations are similar in most important respects.[10.4] They differ materially in one important way. The TeCSA scheme (as it is now known) sets out the purpose of adjudication – that of producing a rapid and economical resolution of disputes – and, while regulation 12 in the statutory Scheme provides that the adjudicator shall act in accordance with relevant provisions of the contract and applicable law, regulation 15 in the TeCSA scheme provides important flexibility. This is that he shall act in accordance with relevant legal rights where this is consistent with a quick and economical dispute resolution machinery, but where this is impossible he shall do so in a way which is fair and reasonable.

Detailed provisions

The important provisions of each scheme are, however, worth summarising.

- Adjudication is commenced by the service of a notice identifying the nature of the dispute in general terms. (Scheme regulation 1; TeCSA regulation 3(i))
- The adjudicator shall be either the person named in the contract or the person nominated by an adjudicator nominating body. (Scheme regulations 2–4; TeCSA regulations 6 and 7)

- The appointment of the adjudicator shall be effected within seven days of the service of the notice. (Scheme regulation 5; TecSA regulation 7) Both schemes contain detailed provisions for the appointment and replacement of adjudicators where the agreed appointee or nominee is unable for whatever reason to act.
- The adjudicator is empowered to act in relation to several disputes in relation to the same contract. (Scheme regulation 8; TecSA regulation 5)
- The adjudicator shall take the lead in establishing the procedure to be adopted. (Scheme regulation 13; TecSA regulation 19). In particular he is empowered to require the delivery of relevant documents and to draw adverse inferences from the failure to supply documents which harm the party's case. He is also empowered to limit the scope of written submissions.
- The adjudicator shall publish his decision within 28 days of his appointment. This period is capable of being extended by 14 days. (Scheme regulation 19; TecSA regulation 22) Under the Scheme this period is capable of further extension but no such equivalent provision exists under the TecSA scheme.
- The adjudicator is empowered under the Scheme (regulation 22) to provide reasons for his decision if requested by one of the parties. Under the TecSA scheme (Regulation 27) he is prohibited from providing reasons.
- The adjudicator has no power to award costs. (Scheme regulation 25; TecSA regulation 24) He does have power to award interest. (Scheme regulation 20 (c); TecSA regulation 26)
- The decision of the adjudicator is capable of summary enforcement. (Scheme regulation 23(2); TecSA regulation 28) The adjudicator's decision is binding upon the parties until the dispute is determined by subsequent arbitration or litigation. The fact that an adjudication has been conducted does not prevent either party from exercising any legal right such as the commencement of arbitration or litigation. Significantly, the reference to litigation or arbitration is not by way of appeal against the adjudicator's decision. Hence, the purpose of the adjudication is to provide an *alternative* mechanism to the traditional ways of resolving disputes.

Perceptions of adjudication

As far as delay claims are concerned, the adjudication procedure does not at first sight appear to lend itself to the sort of complex claims described in Chapter 6. Attempts to have such claims dealt with by adjudication might be thought likely to result in subsequent arbitration or litigation by the party dissatisfied with the adjudicator's decision. To that extent some commentators have suggested that adjudication will simply provide a superfluous layer of proceedings which will fail to achieve any real resolution of the dispute, and furthermore that adjudication is necessarily an inferior way to resolve disputes.

That view finds support in the fact that adjudication is not a new way to resolve disputes. Clause 24 of the standard forms DOM/1 and DOM/2 provided for adjudication in the event of disputes over the right to set off money against interim applications by sub-contractors. The adjudication process was often criticised as being toothless on the grounds that the approach most often adopted by adjudicators was to order the disputed sums to be paid into stakeholder accounts and compelling the dispute to be dealt with by arbitration.

Will the new regime prove any different? Given the approach adopted by the courts, and the increased use of adjudication, this disparagement is a short-sighted view. It also assumes that most of the construction industry actually wants and prefers to see disputes dealt with by way of full-scale litigation or arbitration.

Practicalities

The adjudication machinery may serve a number of practical or tactical objectives. Firstly, the procedural rules provide a distinct advantage to the claiming party. He will be able to take his time in preparing his claim and serve notice when and only when he is ready to bring the claim. His opponent will be constrained by the time limits of the scheme. This will allow him only a period of days in which to respond to a claim which may have taken weeks or months to prepare. (In this respect at least, the critics of adjudication have a point – any system of dispute resolution which manifestly favours one party must be regarded as suspect.)

Secondly, the procedure may allow a claimant to refer specific narrowly defined disputes to adjudication. Examples might include whether a particular notice of non-completion was validly served, whether in particular circumstances liquidated damages were validly deducted, or whether a particular event caused the delays for which one party contends. The adjudication process may therefore serve to narrow the areas of dispute.

Thirdly, while the adjudication does not supplant any of the parties' other rights, it will be obvious that a successful adjudication which can be immediately enforced will certainly boost the negotiating position of that party and will serve to make settlement more rather than less likely. Similarly, the adjudication process may have the effect of indicating the weaknesses in one party's position. Stating the obvious, it makes more sense for both parties for this to be accomplished within the 28 day time-span for adjudication than for the parties to spend months or years in litigation or arbitration.

Each of these points needs to be considered in the light of the provision in the TeCSA scheme to the effect that if the adjudicator considers it impossible to implement the parties' strict legal rights within the context of a quick and economical procedure he shall form a view which is fair and reasonable (regulation 15). The suggestion that the adjudicator should disregard complex legal submissions in favour of a 'common-sense' appraisal of the position may well find favour. Similarly, the claimant who advances a poorly detailed claim may contend that the adjudicator should disregard the guidance offered by the case law and take a broad brush view, concluding that the claim, whatever its shortcomings is bound to have some merit. Alternatively a respondent may suggest that while the adjudication has been brought in relation to specific matters, the adjudicator should look at the 'wider picture' whatever that might be.

Practical experience suggests that this will be less of a problem than was suggested at the time that the Scheme and the TeCSA rules were first published. In fact, the contrary results seem to have been encountered. The tight timetable for adjudications and the interventionist powers granted to them seem increasingly to mean that the adjudicator takes the lead in establishing the relevant facts and legal principles upon which to found his decision. If, therefore, adjudication is to be regarded as a rough and ready approach to dispute resolution, it must be said that adjudicators are giving themselves every opportunity to get to the right answer.

Conclusions

By entering into a contract governed by the Act, the parties are taken to have agreed a method of dispute resolution in which speed and economy are the paramount considerations, and that is their prerogative. It is naive to suggest that a more 'judicial' approach of the sort associated with traditional litigation or arbitration will necessarily be better. Indeed, the mere fact that an adjudication is required to be conducted more quickly than an arbitration does not make it worse. A transparently poor claim will be no better for being presented in the context of an adjudication and the great majority of adjudicators would be unsympathetic to the submission that, because a matter is proceeding by way of adjudication, it need not be persuasively argued or expressed. Indeed experience suggests that the adjudication process will often provide a rigorous examination of the merits or otherwise of a claim. A few examples drawn from practical experience serve to illustrate this:

- An adjudicator will frequently require that the person responsible for the compilation of a claim should attend before him. The adjudicator can and will ask that party to address issues which trouble him or which are inadequately presented.
- Unlike a judge or arbitrator, there is nothing to prevent an adjudicator telephoning a party to attempt to ascertain some relevant fact.
- The right to draw adverse inference from the absence of a particular document or piece of evidence places a heavy onus on the party in question to ensure that his case is properly supported by all the relevant evidence.
- The fact that the adjudicator is not bound by the strict rules of evidence means that in some instances he is entitled to draw inferences to produce a common-sense result, which may differ from that which he would have been achieved had he been bound to rely on formal discharge of the burden of proof.
- The absence, in most cases, of a requirement to give reasons means that an adjudicator can produce his decision safe in the

knowledge that his decision will not be subjected to minute scrutiny by an appellate court.

10.3   Mediation and alternative dispute resolution

The opening section of this chapter dealt with the increasing desire among construction professionals to find a better way to resolve disputes. The growth of mediation and alternative dispute resolution (ADR) of various types can be seen as the response to this wish. While a number of variants exist, the basic premise of all forms of ADR is that the parties appoint a neutral third party to assist them in finding a negotiated solution to their differences. The neutral third party – the mediator – does not impose a result on the parties and has no power to decide on the merits of the dispute although, if the parties wish, he can express a view on the matter.

Mediation as we know it is a relatively recent creation. It has its origins in the mass torts litigation prevalent in the United States of America during the 1970s, particularly the spate of class actions brought on behalf of those allegedly injured by the actions of the pesticide Agent Orange during the Vietnam War, those claiming against tobacco companies and those claiming as a result of personal injury consequent upon faulty silicone breast implants. To a large extent it was the brainchild of David Shapiro, formerly senior partner of the New York and Washington law firm, Dickstein Shapiro Morin and Oshinsky.

The essence of mediation is to enable the parties to find a solution which meets all of their aspirations. The skilled mediator will assist the parties in finding such a solution which need not depend upon strict legal rights but may, where appropriate, allow the parties to come up with a business compromise.

As with adjudication, it is easy to conclude that mediation is not really a suitable means of settling complex delay claims. However, this view might be said to attribute a degree of mystery to delay claims which is not really justified. The advocate in mediation, perhaps more than in any other form of dispute resolution, must be able to explain the essence of his contentions in a way which the mediator can readily understand. He must also have a clear appreciation of the strengths and weaknesses of his case and be able to maximise the impact of the former and explain the latter. Equally, since mediation is a consensual process it depends absolutely upon the willingness of the parties to make it work. Settlement will depend upon the parties being prepared to make compromises.

The comments made in Chapter 9 about targeting the claim to its end user are particularly applicable to the process of mediation. Points must be made in a way which will have the greatest impact on the mediator and the way in which he conveys them to the other party. Naturally, the excessively complex or ill-thought out claim will be harder to explain to the mediator than a claim which is focussed and intelligible. On the basis that mediation is concerned with the structured negotiation of disputes, there is no reason why delay claims cannot be settled by mediation.

APPENDIX ONE
SAMPLE PRELIMINARY CLAUSES DEALING WITH PROGRAMMES

1.0 Programme and progress

Contract period

The Contractor is referred to the details in the Appendix to the Conditions of Contract for the Dates for Possession and Date for Completion.

The Contractor is invited to submit an alternative tender for any shorter Contract Period which he considers achievable.

Additional clause to be inserted if required

2.0 Information to be submitted with tenders

The Contractor shall prepare and submit the following to support each of his tender offers:

a) A Tender Programme which is sufficiently detailed to show clearly the Contractor's intended sequence and method of working. The programme shall include and indicate all major construction activities, the work of all sub-contractors and Statutory Authorities, and all the work covered by prime cost and provisional sums in the Bills of Quantities. The programme shall take full account of normal inclement weather, all holidays and any restrictions to working times and methods provided for in the Bills of Quantities.

The description of the prime cost and provisional sums must be in sufficient detail to enable the requirements of this clause to be met.

b)   A Method Statement fully describing the techniques, plant and equipment to be used by the Contractor in order to achieve the dates depicted in the Tender Programme. The Method Statement will include both the proposed location of major items of plant and temporary works to be executed on site.

c) (i)   A schedule of Information Requirements to complete the project, stating the date by which each item of information is required. This schedule must allow for the progressive release of information related to the Tender Programme and must take account of the procurement times of construction materials, plant and equipment.

Clause C(i)-(iii)

Alternative clauses as required to be agreed with Design Team and Client.

c) (ii)   Appendix__ indicates in summary terms when the information will be released. The Contractor may not assume at tender stage that the information can be issued earlier than stated within the schedule.

c) (iii) Should the requirements of the Contractor's own programme demand earlier release of information then these items must be separately identified within the tender submission.

d)   A chart indicating the Contract Management Structure demonstrating the responsibilities of on-site and off-site staffing. Attached to this chart should be the names of the members of staff who will undertake the duties indicated.

This clause to be included in addition to any other preliminaries requirement. It may be necessary to also request details of the duration and percentage commitment of each member of the Contractor's staff.

The Contractor's attention is drawn to the fact that all the above information to be submitted with the tender will be discussed at an interview with the Contractor prior to his appointment.

Appendix One

3.0 Information to be submitted after acceptance of tender

Include the sum £........ for providing the service
described in the following clauses 3.1 to 3.6. This sum will
be reimbursed to the contractor in the following stages:

a)   When the Architect notifies the Contractor in writ-
     ing that he is satisfied with the network and detail
     of the Construction Programme, and the support-
     ing information referred to in the following clauses
     3.1 to 3.3 the amount of £......... will be included
     in the certificate following this notification.

It is suggested that the figure inserted in the cash column should be calculated by reference to programming costs included in a building contract of comparable character, size and duration and should not exceed 0.5% of the anticipated tender figure.

b)   The Contractor will be reimbursed a figure not
     exceeding £......... in equal monthly instalments
     until practical completion of the project provided
     that the Architect is satisfied that the requirements
     of the following clauses 3.4 to 3.6 are diligently and
     regularly fulfilled.

The aggregate of the monthly instalments during the duration of the contract referred to in this clause should follow the approximate cost of a reasonable programming service (to avoid any claims for excessive reimbursements in the event of extensions of time being granted) with the balance of monies up to the sum included in the cash column being included in clause a) above.

3.1 Construction programme

So soon as is possible after the execution of the Contract,
and in any event by four weeks after starting on site, the
Contractor is to prepare and submit a detailed Construction Programme in the form of a critical path network, in
accordance with the following criteria:

Dependent on timing and nature of contract, this clause may require amendment.

a)   The programme must clearly establish the logical sequence of work, display any programme restraints determined by resource limitations, and establish all critical activities and their relationship with all other major activities.

(b)   The programme shall include and indicate clearly all major construction activities, the work of all subcontractors and statutory authorities, and all work covered by prime cost and provisional sums in the Bills of Quantities. The programme shall take full account of normal inclement weather, all holidays and any restrictions to working items and methods provided for in the Bills of Quantities.

c)   The programme shall be in the form of a network, either precedence or activity on arrow, and be analysed by computer using a recognised Critical Path Analysis software package.

d)   The network is to be in sufficient detail to satisfy the Architect or his representative that it demonstrates the logical relationship between all construction activities.

The term 'Architect' may require amendment to ensure consistency with the rest of the contract.

e)   Each activity on the network diagram shall be so annotated as to identify its location, trade and fix; and, if so requested by the Architect or his representative, the resource levels required to achieve completion within the duration indicated. The Contractor is to provide to the Architect neatly drawn copies of the network and a copy of all data used in the analysis of the network in the form of a floppy disc in MS-DOS format, compatible with and usable on a pc running (for example) Windows 98.

The purpose of specifying the equipment is to establish the minimum requirements for compatibility rather than determine the physical nature of the hardware.

3.2 Master Programme

The Master Programme for the works will be based upon and entirely compatible with the Construction Programme. All revisions to the Master Programme, made in accordance with the Conditions of Contract, shall be accompanied by sufficient data to demonstrate clearly the logic of such amendments.

This paragraph should be included for JCT 98 Form of Contract but will require amendment to suit other Contracts.

3.3 Supporting information

The Contractor is to provide, with the Construction Programme, the following supporting information compatible with the programme.

a) An update of the Method Statement provided with his tender describing fully the techniques, plant and equipment to be used.

b) An update of the Schedule of Information Requirements itemising all construction information and the dates by which each item is required to achieve the programme.

c) A Procurement Schedule itemising major construction materials and subcontractors, giving dates by which all stages of the procurement process, including sub-contractor's drawing production, must be achieved.

3.4 Detailed programme

The Construction Programme information shall be expanded during the course of the project with the following information, two copies of which must be supplied to the Architect:

a) Short-term programmes, covering a period of say 2–3 months, which need not be in the form of a network although they must relate directly to activities within the Construction Programme.

b) Sub-contractor's programmes indicating all off-site activities including the production of drawings and details. Due allowance must be made within these programmes for submission of drawings to the Architect for comment; a two week period for the Architect to examine, mark up and return the drawings with comment; and their subsequent amendment by the sub-contractor to take account of any comments received.

Subject to amendment to achieve compliance with Architect's brief.

3.5   Generally

The networks and bar charts are to be prepared and monitored by a suitably qualified person who is to remain in close contact with the site until contract completion.

One copy of all current programmes and schedules are to be available on the site at all times.

Submission of these programmes and schedules will not relieve the Contractor of his responsibilities under the Contract.

3.6   Monitoring

Immediately prior to each Site Meeting the Contractor is to meet with the Architect or his representative to agree the actual progress of the Works.

The term Architect may require amendment to ensure consistency with rest of contract.

The Contractor is to update the Construction Programme network to take account of progress of all current activities as of the date of the progress meeting. This update is to form the basis of the Contractor's progress report for the Site Meeting.

A copy of the updated network computer data is to be provided to the Architect in the form of a floppy disc in MS-DOS format, compatible with and useable on a pc running (for example) Windows 98.

The term Architect may require amendment to ensure consistency with rest of contract.

A written statement of Information Required by the Contractor is to be issued at each progress meeting.

Subsequent to the production of the progress report, the Contractor is to update the network in order to demonstrate, where possible, how he intends to overcome any delays which may have occurred. The changes in logic and/or durations will be submitted to the Architect or his representative in the form of a schedule indicating both the previous and amended data, together with amended network diagrams.

The term Architect may require amendment to ensure consistency with rest of contract.

In addition, where the sequence of work on site is varied, the Contractor is to revise the Construction Programme prior to the next monthly update to reflect such amendments.

APPENDIX TWO
DRAFT NOTICES OF DELAY

(1)   Dear Sir

     [Heading]

     We are writing pursuant to the requirement of Clause 25.2.1 of the contract between our respective companies to notify you of the following event:

- instructions to replace timber window frames with UPVC window frames

     which we believe is likely to delay the progress of the works because

- these components are on a 6 week delivery period; and
- installation of these components is schedule to take place 4 weeks from today

     We believe that this will delay the progress of the works as a whole because this work is essential to achieving water-tightness of the building. Hence, we consider this to be a relevant event for the purpose of Clause 25.4.5.1 of the contract. Our present estimate of the extent of the delay is 2 weeks.

     Yours faithfully

(2)   Dear Sir

     [Heading]

     Because of the instructions to omit timber windows and replace them with UPVC windows, our works are likely to be delayed, and we anticipate that since we are scheduled to instal the windows in 4 weeks time, even though the UPVC windows are on a 6 week delivery, this delay will be 2 weeks.

     Yours faithfully

NOTES

Chapter 1:

1.1   Aside from the fact that canal ('navigation') construction gave the world the word 'navvy', the story of the canal age provides a fascinating insight into the problems which the pioneers of modern civil engineering encountered. Ground conditions, changes in the works, bewilderingly incompetent design and supervision – it will all be very familiar, and perhaps the only significant differences were the risk of the workforce being rendered incapable by typhoid and the threat of Napoleonic invasion.

1.2   The debate in the construction press reached heights of intensity seldom matched in recent years. Whether the sometimes inflammatory comments are justified is dealt with elsewhere, but it must be said that it is still relatively early days and, to paraphrase Mikhail Gorbachev, anything which encourages, 'thinking in new ways' is surely healthy.

Chapter 2:

2.1   It is worth comparing the respective obligations imposed by clause 2.5.1 of WCD 98 and those which the Architect's Appointment (used in conjunction with the traditional contract) provides. Commentators agree that the reference in clause 2.5.1 to an architect is intended to impose the obligation to exercise reasonable skill and care in the design of the works. The important distinction is that under the traditional contract, where the architect is the appointee of the employer, the extension of time and loss and expense clauses entitle the contractor to relief when problems occur for which the architect is as fault. By contrast, under the design and build form, the design of the works is the responsibility of the contractor. It should also be noted that this particular clause has been the cause of vast amounts of time and effort being expended by the draftsmen of contract amendments and warranties in attempts by employers to increase the risks borne by contractors.

2.2 Paper presented to the King's College Centre for Construction Law and Management by Richard Winward.

2.3 Report of the 'Committee on the Placing and Management of Contracts for Building and Civil Engineering Works' chaired by Sir Harold Banwell. In thirty years the wheel has come full circle, Sir Michael Latham attributes many of the industry's ills to problems caused by over-elaborate and confrontational contract documents.

2.4 In paragraph 3.7. Latham discusses the creation of a contract strategy determining the level of risk which the employer wishes to undertake. He identifies three separate types of risk: fundamental risks – war damage, nuclear pollution and supersonic bangs; pure and particular risks – the former covering fire damage and storms, the latter collapse, subsidence vibration and removal of support; and speculative risks – ground conditions, inflation, weather, shortages and taxes. This of course is sensible but it does not really address the principal question which depends upon determining the way in which the particular works can best be constructed.

2.5 Again it is interesting to note the reaction of the Latham Report, and particularly the comment at paragraph 2.4 that his brief did not involve consideration of the economic climate in which his report was produced. The difficulty then is that unless the marketplace permits the good practices which he espouses to be implemented economically, there is the risk that the industry will be unable to afford to follow his recommendations.

2.6 This general statement has been slightly blurred by the decision of the Court of Appeal in *Crown House Engineering Ltd* v. *AMEC Projects Ltd* (1989) 48 BLR 32. That case contains what can only really be described as unhelpful and apparently off-the-cuff suggestions that, despite the absence of a contract, it was possible for the parties to rely upon timing obligations which would have applied had a contract come into being. It is uncertain whether any weight can or should be attached to these observations. Perhaps the sensible and cautious view is that the relevant passage in the judgment of Lord Justice Slade should be treated as dealing with the facts in that particular case, and not as having any wider application.

2.7 In this respect Sir Michael Latham takes an almost directly contrary view. At paragraph 5.21, the report suggests what amounts to a legislative ban on 'bespoke' amendments to standard form contracts. The view seems to be that disputes are caused by onerous or one sided contracts and that this can be cured by outlawing provisions of this type. This is the corollary to the view which appears in paragraph 5.17 which recommends among other things, the imposition of a general duty to trade fairly, while indicating that the approach of the New Engineering Contract seemed very attractive. Two brief comments are sufficient; firstly that fairness is not a quality which can be imposed – in the context of a competitive market, fairness is not the first concern of any party, and it is unrealistic to think otherwise; secondly that when applied to the NEC, it is easy to confuse simplicity and fairness – the criticism most frequently aimed at the NEC is that the language of the contract embraces simplicity sometimes at the expense of clarity or meaning.

2.8 Particular thanks are due to David Mace of Johnson Jackson Jeffs for allowing me to reproduce some sample contract clauses produced by his firm.

2.9 In this respect, it is relevant again to consider the fact that the Latham Report adopts an equivocal approach. At paragraph 2.4., its author admits that adivising on government policy plays no part in the committee's brief, while conceding in paragraph 2.5 and 2.6 that the construction industry has been hit worse than most by the recession, and acknowledging that the industry's recovery depends upon government economic policy. The conundrum which this reveals is that while calling for the development of good trade practices, the report accepts, albeit tacitly that the prospects of those being developed is significantly lower if members of the industry are seriously stretched in financial terms.

2.10 It is worth observing that the suggestion that a construction project is a competitive match will probably find little support either with Sir Michael Latham or with the draftsman of the NEC. That, it is suggested, is to ignore the fact that construction projects are undertaken by parties both of whom wish to make money from the venture. Even where one of them is a public authority, the use of public money and the growth of Compulsory Competitive Tendering means that the prime consideration in both public and commercial projects will be the budget. Thus, it is possible for all parties to be winners. Equally, both can be losers. Achieving the former and avoiding the latter is, at least in this writer's view, a product of certainty and developing good habits rather more than the elusive concept of good faith.

2.11 Lord Pearson in *Trollope & Colls Ltd* v. *North West Metropolitan Hospital Board* [1973] 1 WLR 601. The importance of this case is to provide an example of the principle that the courts will not improve upon a bad or even a silly contract in circumstances where the express terms of the contract were clear.

2.12 Lord Cross of Chelsea in *Liverpool City Council* v. *Irwin* [1977] AC 239 at page 262. Here too it bears repeating that the terms which will be implied will only be those necessary to make the contract *work* rather than making it ideal for one or other of the parties.

2.13 *Luxor (Eastbourne) Ltd* v. *Cooper* [1941] AC 108.

2.14 Lord Cross in *Liverpool City Council* v. *Irwin*.

2.15 Sir Anthony May, *Keating on Building Contracts*, 6th Edition, Sweet & Maxwell.

2.16 *Liquidated Damages and Extensions of Time* by Brian Eggleston. Published by Blackwell Science 1997.

Chapter 3:

3.1 Inevitably this is a view that clashes with some commentators. It is supported by both observation of major projects which have gone seriously wrong and advising on others while attempting to apply the lessons taught.

3.2 As part of the Mediation Workshop organised by the Centre for Dispute Resolution.

3.3 4th edition, published 1978. Interestingly, although the present 6th edition makes the same point – in effect that the architect owes an independent duty to consider delays and extensions irrespective of compliance by the contractor – the learned editor suggests that this may perhaps not be applicable to the 1980 version of the JCT form. Unfortunately, he does not explain why he considers that clause 25.2 differs in meaning from its forerunner. Certainly, it is difficult to see anything in the actual wording which would justify this view. The position is unchanged with regard to JCT 98.

3.4 In conversation with the writer, July 1995.

Chapter 4:

4.1 The position is obviously slightly different where the contract provides for sectional completion.

4.2 Except where the context demands, the expression 'Practical Completion' has been used throughout.

4.3 The question of disruption and acceleration is a vexed issue which is dealt with unsatisfactorily in the JCT and ICE forms. It is approached rather more comprehensively in the NEC. These issues are briefly considered at the end of this section.

4.4 (1) In contracts providing for sectional completion the position is modified in that the notice or certificate is issued upon non-completion of a section of the works and damages are payable at the rate for the sections prescribed in the contract. Brian Eggleston also provides an extremely useful summary of the relevant provisions of the relevant standard forms.

(2) The conventional wisdom is that entitlements to liquidated damages and additional time are opposite sides of the same coin. It is possible with a little imagination to envisage circumstances where completion of the works is geared to a particular event and hence, even if completion has been delayed by an event which would otherwise have entitled the contractor to additional time, the fact that this milestone had been missed might be said to give rise to an entitlement for the employer to claim liquidated damages. The problem with such an argument is that it would almost certainly fall foul of the rule that the party relying upon the delay should not be able to rely upon his own breach (see for example *Amalgamated Building Contractors* v. *Waltham Holy Cross UDC* [1952] 2 All ER 452).

Chapter 5:

5.1 Both these cases are discussed in detail in Chapter 6.

5.2 The use of claims as a weapon to enforce bargaining strength has been most forcefully attacked by a number of articles in *Building* magazine, by Rudi Klein of the Specialist Engineering Contractors Confederation, an organisation whose objects include the promotion of the interests of specialist sub-contractors.

5.3 For a more detailed consideration of what partnering actually means both in theory and practice, see Julian Critchlow's *Project Partnering in the Construction Industry*.

5.4   *Holme* v. *Guppy* and *Roberts* v. *Bury Commissioners* provide examples of the former. Modern cases where the extension of time machinery has broken down have tended to turn upon very particular facts and, it is suggested provide little in the way of general guidance. Of those cases, that most frequently referred to, *British Steel Corporation* v. *Cleveland Bridge* 24 BLR 97 is to treated with care because it is not a case in which the timing machinery had broken down; rather it is a case where no time for completion had been agreed. Of the remaining cases, *Peak* v. *McKinney*, provides what, at best is flimsy support for the proposition that an extension of time machinery may break down leaving time at large. In the most recent, *McAlpine Humberoak* v. *MacDermott* the Court of Appeal gave short shrift to the judge's original finding that time could be said to be placed at large.

5.5   There is a clear and thorough analysis of this point in the Australian case *SMK Cabinets* v. *Hili Modern Electrics Pty Ltd* [1984] VR 391 by Mr Justice Brooking. In that case there was no extension of time clause. See also *Peak* v. *McKinney* (1970) 1 BLR 114 where this is the effect of the Court of Appeal's judgment.

5.6   See Section 2.5. In that case, evidence was called by the defendant to the effect that the contract had been deliberately drafted in a manner which placed the risk of delaying or disruptive events upon the plaintiff sub-contractor. As discussed above, the case concerned an attempt on the part of the plaintiff to construe the contract in a way which relieved the plaintiff of this risk.

5.7   See in particular Section 2.2.

5.8   In this context the word 'correspondence' draws no distinction between letters, memoranda, notices, statements, record sheets, faxes, e-mail or any other medium where the written word is passed from one person to another and some means exist for preserving it.

5.9   Keith Pickavance *Delay and Disruption in Construction Contracts*, LLP, 1997.

5.10   *Wraight* v. *PH&T Holdings* (1968) 13 BLR 26.

5.11   Despite the title of his book, Keith Pickavance does not really deal with disruption at all save to define it in rather cryptic terms as 'An adverse effect on the progress of the works. Since it does not affect completion it can never be described as excusable.'

5.12   It is suggested that the contrast with claims for delays can be summarised as follows:

- To succeed in proving delays, the contractor must identify a particular matter which led to more time being required in order to complete the works.
- To succeed in a disruption claim, the contractor needs to prove that the event complained of *did* cause the need for additional resource, and, in most cases, that his own planning and management of the job were not in fact the real causes.

Chapter 6:

6.1 Research by Julian Critchlow (see 5.3 above) suggests emphatically that this comment can be applied equally to arbitration.

6.2 These points formed the central theme of the paper given by Dr Martin Barnes at the SBIM conference on the New Engineering Contract on 3 February 1995. Dr Barnes, who headed the drafting committee of the NEC, advanced a broadly similar argument in opposing the motion 'This House believes that the NEC does not meet the foreseeable needs of the construction industry' debated by the Society of Construction Law on 7 March 1995 (The motion was defeated by the narrow margin of 53 votes to 49 with 4 abstentions.)

6.3 This part of the text is an expanded and up-dated version of the author's '*Wharf Properties* v. *Eric Cumine Associates*. The effect on rolled up claims.' [1991] 7 CLJ 303.

6.4 An example comes in Brian Eggleston's *Liquidated Damages and Extensions of Time* where he suggests that the court actually struck out the statement of claim as disclosing no reasonable cause of action. This was not the case. The action was struck out because it was held that the way in which it was pleaded was embarrassing to the defendants and prejudicial to their chances of a fair trial. This apparent confusion seems to carry over into Mr Eggleston's comments on the *Devonald Williams* case which he describes as holding that a rolled up claim for an extension of time could be pursued if it could be shown that it was not feasible to put it any other way. As will be apparent from page 121 this is not a view I share, although for the reasons set out there the uncertainty seems to have been shared by the judge in that case.

6.5 I apologise for the fact that a certain amount of legal jargon is unavoidable in describing the points analysed in this case. I have attempted where necessary to explain the importance of the concepts involved. I have not however done this in relation to the procedural ebbs and flows which characterised this case which can largely be taken as read.

6.6   This observation is implicit in the commentary of the editors of *Construction Industry Law Letter* shortly after the decision in *Wharf* was reported.

6.7   This procedure known as the 'case stated' is no longer available. The result of it in both of these cases was somewhat akin to setting an exam paper for the judges. All of the commentators agree that both would have passed the test with honours.

6.8   See discussion of *ICI* v. *Bovis* on page 126.

6.9   Even if *Wharf* was not actually cited to the court, there is little doubt that it figured prominently in the defendant's thinking.

6.10   February 1995.

Chapter 7:

7.1   Obviously there is no data dealing with the split between 'good' and 'bad' claims which proceed to hearings. Experience suggests that there is a preponderance of the latter.

7.2   'An extreme example of this from the 19th century that probably could not be repeated today is how the Board of the Great Western Railway implemented the decision to change from the broad gauge with which it was originally built to the standard gauge of the rest of the county. To minimise the inconvenience to freight and passengers the work was done over a single weekend with meticulous planning and preparation and separate gangs of labourers employed to work on every few miles of track.'

7.3   *Critical Path Analysis, and other project network techniques* by Keith Lockyer and James Gordon, Financial Times (1991) Prentice Hall Paperback.

7.4   The expression 'critical path analysis' is used widely, and often inaccurately and for this reason it is an expression I have used sparingly – in this section it is used only in its strictest sense.

7.5   It is not the only method of network analysis. I have concentrated upon it because it is the method most easily adapted to construction projects. Additionally, because the purpose of this work is to examine the effects of delays upon construction work, and because once one technique is appreciated the others are not difficult to grasp, I have not attempted to consider (for example) activity on node methods of programming.

7.6   A good example of this principle would be the wish to see the Jubilee Line Extension completed by the Millennium.

7.7   In most instances this exercise is done by taking off the time periods from a bar chart and applying resource output levels to the time required, or in reverse by applying the resource output levels to the available resources to calculate the time which will be required. In more complex projects this exercise will be extended to apply it to the dependencies shown on the draft network to determine whether there are factors which will limit the available resources. It may perhaps be the case that the finish date for a preceding activity which uses some or all of the same manpower will limit the availability of manpower for this or succeeding activities.

7.8   What is clear from the writer's experience is that the inadequacy of the programme or the imperfect understanding of its operation meant that for long periods of such projects the works proceed on a hand to mouth basis without any real programme at all – the parties simply got on with the available work as best they could.

Chapter 8:

8.1   One of the particularly rewarding features of Keith Pickavance's *Delay and Disruption in Construction Contracts* is his discussion of the way in which the analysis of delay has been approached in many decisions of the various Boards of Contract Appeals in the United States. He makes the valid and important point that while the reasoning of the American courts is published in the reports of the Boards of Appeals, this is not the case in Great Britain where most disputes are dealt with by arbitration which takes place in private. It is difficult to dispute his view that this has led to this country lagging behind in the development of delay analysis.

8.2   Possible exceptions to this may occur where there is debate as to whether the resources extracted from the tender were actually capable of producing the levels of output contended for, or whether particular programmes activities were capable of performance within particular time periods.

8.3   *Building Contract Litigation* 4th Edition 1993 Longman – interestingly this is not the view expressed in the first two editions of the work. It is suggested that the later view is to be preferred.

8.4 Both cases are considered in detail in Chapter 3.

8.5 This argument has been encountered in a number of highly innovative contracts where the contractor has said quite openly that, because no job of precisely similar type had been undertaken previously, aspects of the construction process represented a best guess as to the labour or resource requirements which might be needed.

8.6 This is a comment based on the practical realities of most claims. Although it may be possible to produce an analysis of the theoretical effect of particular events, this approach will not be possible if the tender is unsifficiently detailed to allow this type of analysis.

8.7 The attraction of the 'aggregate delays less mitigating events' approach adopted by a number of construction claims consultants is that it can with relatively little effort produce an entitlement to the whole of the period of delay which can be attributed to the other party. Its problem is that it is a completely useless method of analysing the delays which were actually experienced.

8.8 The former view is expressed by Tony Farrow of Trett Contract Services in Issue 20 of *Trett Digest*. The latter view is put forward by a number of correspondents to the following issue. It follows from the comments made in the remainder of this section that both views may result from investing the concept of float with a meaning which it may not have.

Chapter 9:

9.1 A useful adage is 'tell them what you are going to tell them, tell them, and then tell them what you have just told them'.

9.2 The decided cases provide very little real assistance. Invariably, they depend on our individual facts. Except where it can be shown that the architect or engineer has required levels of detail so extreme that he cannot be said to be acting in good faith, this argument will be difficult to prove.

9.3 An application in 1995 by a body called the Institute of Commercial Litigators to gain High Court rights of advocacy for claims consultants met with little favour from the Lord Chancellor's department. The report of the Lord Chancellor's department can be read as an explanation of some of the difficulties which may be encountered when claims consultants conduct arbitration.

9.4   Most notably *James Longley & Co Ltd* v. *South West Thames Regional Health Authority* (1983) 25 BLR 56 and *Piper Double Glazing* v. *D C Contracts* (1992) 64 BLR 32.

9.5   See *New Victoria Hospital* v. *Ryan* (Unreported)

9.6   Interestingly the editors of the *Construction Industry Law Letter* described Mr Justice Laddie's approach as reactionary. In the light of Part 35 of the Civil Procedure Rules, it seems unlikely that they would continue to hold this view today.

Chapter 10:

10.1   Both documents appear to regard contract forms as being the cause of conflict. It follows from Chapter 2 that this is to confuse uncertainty over what is to be constructed and the allocation of risk with the form of contract selected to record the agreement of the parties. Latham's approach is to recommend the use of 'simpler' contracts such as the NEC. Egan's more radical approach is to recommend the adoption of non-contractual relationships as a way to avoid the development of conflict.

10.2   The report of the symposium which appeared in *Building* magazine does not really do justice to the surprising degree of consensus which was achieved by a panel drawn from a wide cross-section of the industry.

10.3   Now the Technology and Construction Court Solicitors Association (TeCSA), a body comprised of solicitors regularly practising in the Technology and Construction Court (formerly the Official Referee's Court), the division of the High Court dealing with construction disputes.

10.4   It has been suggested that the drafting of the Scheme leaves something to be desired. It remains to be seen whether these criticisms have substance. It is certainly the case that the ORSA regulations are rather more user-friendly and easier to assimilate.

TABLE OF CASES

233

234

Table of Cases

BIBLIOGRAPHY

Bartlett, A. (ed.) *Emden's Construction Law*, Butterworths.

British Standard 5750/ISO 9000, published by The Stationery Office.

Chartered Institute of Building *Guide to Critical Path Networks*, CIOB Publications.

Cornes, D. (1997) *Design Liability in the Construction Industry*, 4th edn, Blackwell Science.

Critchlow, J. (1996) *Project Partnering in the Construction Industry*, LLP.

Collins, L. et al (eds) (1993) *Dicey & Morris on the Conflict of Laws*, 12th edn, Sweet & Maxwell.

Department of the Environment (1994) *Constructing the Team* (the Latham Report), The Stationery Office.

Duncan Wallace, I.N. *Hudson's Building and Civil Engineering Contracts*, 10th edn, Sweet & Maxwell.

Duncan Wallace, (1994) I.N. *Hudson's Building and Civil Engineering Contracts*, 11th edn, Sweet & Maxwell.

Eggleston, B. (1997) *Liquidated Damages and Extensions of Time*, 2nd edn, Blackwell Science.

Fenwick Elliot, R. (1993) *Building Contract Litigation*, Longman.

Keating, Donald, *Building Contracts*, 4th edn, Sweet & Maxwell.

Keating, Donald, *Building Contracts*, 5th edn, Sweet & Maxwell.

Lockyer, K. and Gordon, P. (1991) *Critical Path Analysis and Other Project Planning Techniques*, Financial Times/Prentice Hall.

May, Sir A.(1995) *Keating on Building Contracts*, 6th edn, Sweet & Maxwell.

National Economic Development Office, *Achieving Quality on Building Sites*.

Nisbet, J. *Fair and Reasonable Building Contracts Since 1550*.

Pickavance, K. (1997) *Delay and Disruption in Construction Contracts*, Informa Publishing Group.

Proceedings of the King's College Centre for Construction Law and Management.

Report of the Committee on the Placing and Management of Contracts for Building and Civil Engineering Works chaired by Sir Harold Banwell.

INDEX